AF884919

SOTHIS PRESS

Sothis Press aims to completely reimagine the public domain book with an emphasis on universal access to captivatingly designed print editions. We believe that technology will enable every single book ever published to be perpetually accessible in beautiful print form. Sothis is beauty in ubiquity.

www.sothis.press
1 Waterhouse Square, London, ECIN 2ST

The Chemistry of Paints and Painting

Sir Arthur H. Church

SOTHIS PRESS

ISBN 978-1-0365-0842-5

The Chemistry of Paints and Painting
is in the public domain.

This edition is published by
Sothis Press © 2023. The company
is registered in England and Wales.
The Company Number is 14900389.

C E Young

THE

CHEMISTRY

OF

PAINTS AND PAINTING

BY

A. H. CHURCH, M.A., F.R.S.

Professor of Chemistry in the Royal Academy of Arts in London

SECOND EDITION

REVISED AND ENLARGED

LONDON

SEELEY AND CO., LIMITED

ESSEX STREET, STRAND

1892

TO

SIR FREDERIC LEIGHTON, BART., P.R.A.,

WHO HAS ALWAYS SHOWN A DEEP INTEREST

IN ITS SUBJECT, AND HAS GREATLY

ENCOURAGED ITS AUTHOR,

THIS BOOK

IS, BY PERMISSION, DEDICATED.

PREFACE TO SECOND EDITION

In preparing for the press a second edition of this book so soon after the publication of the first, I have not made many corrections or additions. Thanks, however, to the suggestions of correspondents and critics, some new results have been incorporated with the work, and some omissions supplied. To Mr. A. P. Laurie, of King's College, Cambridge, I am indebted for interesting facts concerning the behaviour of certain pigments. My own researches, on many of the questions connected with the very wide range of subjects considered in this handbook, are being pursued so far as time and opportunity permit.

A. H. C.

PREFACE

In the present volume the materials and methods of the painter of pictures are viewed mainly from the chemical standpoint. An attempt has been made to treat in orderly sequence the various kinds of painting-grounds, the constituents of vehicles and varnishes, the pigments themselves, and the chief processes of painting. Although the artistic side of the numerous problems discussed has not been neglected, the book is in no way intended to teach manipulation to art students. It has been written with the view of explaining to artists, whether they be accomplished masters or commencing students, the chief chemical and physical characters of the materials with which they deal and of the operations they practise. In many instances a sketch of the processes for preparing certain pigments and varnishes is given, not in order to turn the painter into a colour-maker or a varnish-manufacturer, but rather that he may acquire a clearer insight into the nature and properties of the most important constituents entering into the composition of his pictures. With regard, however, to the tests for purity and genuineness which I have described in the following pages, my object in introducing them has been different, for I trust that (in some cases, at least) the easy experiments I have recorded will be tried, especially with suspicious pigments. The operations require but little time; the pieces of apparatus needed, like the chemicals, are few and inexpensive. And when the ease with which these testings can be made has been proved by practice, the experimenter may perhaps be induced to proceed a little further, preparing his own siccative oil, selecting and purifying his spirits of turpentine, and so forth. That the painter should test the varnishes he buys for hardness and toughness, and the pigments for durability, may, I hope, be taken for granted.

'Titian managed pretty well without chemistry, did he not?' A distinguished artist asked me this question the other day. But not only were the conditions under which the painters of Titian's time worked simpler than those of the nineteenth century, but grounds, paints, oils, and varnishes were generally prepared in the studios of the artists, and under their own superintendence, so that the chances of going wrong were comparatively limited. And it is not to be denied that a better acquaintance with the nature of the materials which many of the old masters employed would have caused their works to be handed down in sounder preservation to future generations.

It is possible—I hope, indeed, it is probable—that this book may be found of service to students who are purposing to devote themselves to certain manufacturing and technical pursuits. I am aware that to those who refer to its pages for the revelation of all the secrets of colour-manufacturers it may prove in some measure disappointing; yet I trust that, in the way of information and suggestion, the study of this volume will not be unattended with advantage. It must be remembered that it is confessedly an elementary manual only, written with a definite aim, but covering a very wide area of inquiry. And if chemists should conclude that it contains too little chemistry, artists may perhaps think that it contains too much.

There are repetitions in the following pages, for the topics discussed in some of the chapters overlap one another. I am perfectly aware of having made the same statement, given the same figure, and expressed the same opinion in more than one place. The scheme of the work required such repetitions. I felt sure that many an artist or student would turn to one section or other of the book without caring to read the whole. One inquirer would like to ascertain at once what pigments were safe, what dubious, what fugitive, by a reference to the tables in Chapters XXI. and XXII.; while another, anxious to learn something of the evidence on which the several verdicts of approval or condemnation were based, would expect to find his requirements met in the pages devoted to trials of pigments. Again, under the names of the individual pigments, discussed in Chapters XIII. to XIX., some of the changes described in the last part of the work are quoted. Thus it happens that there are some materials common to all of those sections of the book just named.

Much of the substance of the lectures which I have

delivered before the Royal Academy since the year 1880, has been incorporated with the present manual. But it is necessary to state that some of the original material to be found in the following pages has been long before the artistic world, and has found its way into the books and essays of other writers. I say this, not for the purpose of making reclamations of priority, but in order to prevent myself from being charged with plagiarism. For instance, so long ago as 1859, I described, for the first time, some of the artistic uses of solid paraffin in a paper on the processes of painting, read before the Oxford Architectural Society; further details were given in a lecture to the Architectural Association in 1862. On many other matters connected with the chemistry of paints and painting, new investigations and studies were published by me between the years 1867 and 1872, particularly in notes and essays entitled 'Chemical Aids to Art,' and 'The Chemistry of the Fine Arts.' But my statements and results, whether contained in the above publications or in my Academy lectures, have not been, in all instances, referred to their source, or reproduced with accuracy, while some have been overlooked or forgotten.

In preparing the present volume I have made considerable use of several of the works named in my *Bibliographical Notes*; I have consulted also the standard chemical dictionaries of Watts and of Wurtz, the treatise by Roscoe and Schorlemmer, besides many special papers by other chemists. I wish I could have given an authority for every statement not derived from my personal experience, but, in an elementary manual treating of many diverse topics, such a plan, even if it could have been carried out, would have embarrassed my story with a multitude of perplexing references.

I do not know of any one text-book which covers the same ground as the volume now offered to the public. Several small books on pigments—the most important of all the materials employed by the artist—have indeed been lately published, but the chemical information they afford is generally meagre, and sometimes far from exact. One recent little brochure, which lies before me, has, I confess, caused me some amusement not wholly unshaded with regret. The writer does not pose as a humourist, yet he tells us, when we test for lead in cadmium red, first to mix the sample with white lead before applying the usual test for that metal. Chinese vermilion, he informs us, is sulphide of *arsenic*, though it is really sulphide of mercury. The presence of *sulphides* of baryta and lime is stated in one place to lend a softness to the chromates of lead: as these sulphides instantly blacken these brilliant chromates, perhaps they may be said to soften them. Cœruleum, a stannate of cobalt, is directed to be made of carbonate of soda, powdered flint, and oxide of copper, its two essential constituents, the oxides of tin and cobalt, not being named. These and many other equally preposterous statements and directions may afford merriment to the chemist, but it is indeed pitiable that such teaching should be seriously offered to artists and art-students.

It is satisfactory to know that several accomplished

chemists are now devoting themselves to the practical study and improvement of pigments. Mr. A. P. Laurie, Mr. H. Seward, and Mr. J. Scott Taylor, are all doing good work in this direction.

It remains for me to express the hope that the readers of this volume will favour me with any material at their disposal which may serve for the correction and improvement of its pages. I am aware of having omitted to notice many interesting matters; amongst these I include certain pigments, derived from coal-tar products, which have not yet been sufficiently tested. Then, too, the materials and methods of ceramic and glass painting have been excluded from consideration, mainly because their adequate treatment, while demanding much space, would have appealed to a comparatively limited group of students.

If painters and chemists will grant me their help, I trust that I may further justify, by means of an improved edition of my book, the favourable reception which I hope may be accorded to the first.

A. H. CHURCH.

KEW, *March*, 1890.

BIBLIOGRAPHICAL NOTES

Andés, L. E., 'Oel und Buchdruckfarben.' Pp. viii., 271. Leipzig, 1889.

Andres, E., 'Volatile and Fat Varnishes,' Brannt's translation. Pp. xx., 346. Philadelphia, 1882.

Blockz, J., 'Peinture à l'Huile.' Pp. iv., 98. Gand, 1881.

Cennini, Cennino, 'A Treatise on Painting in Fresco, Secco, Oil and Distemper.' Completed in 1437; first published (in Italian) in 1821; translated into English by Mrs. Mary P. Merrifield. Pp. lxx., 177. London, 1844.

Eastlake, Sir C. L., 'Materials for a History of Oil-Painting,' Series I., pp. xii., 561; Series II., pp. xv., 432. London, 1847, 1869.

Field, G., 'Chromatography, a Treatise on Colours and Pigments.' Pp. xix., 276. London, 1835.

Field, G., 'Chromatography,' edited by T. W. Salter. Pp. vii., 436. London, N.D.

Field, G., 'Chromatography,' modernized by J. S. Taylor. Pp. viii., 207, 2nd edit. London, 1885.

Gentelé, J. G., 'Lehrbuch der Farben-Fabrikation,' 2te Auflage. Braunschweig, 1880.

Lefort, J., 'Chimie des Couleurs.' Pp. 344. Paris, 1855.

Merrifield, Mrs. M. P., 'Original Treatises (of the twelfth to the eighteenth century) on the Arts of Painting,' with Introduction and Notes. Two vols.; pp. cccxii., 918. London, 1849.

Mierzinski, S., 'Die Erd- Mineral- und Lackfarben,' 4te Auflage. Pp. viii., 500. Weimar, 1881.

Riffault, Vergnaud and Toussaint, 'A Practical Treatise on the Manufacture of Colours for Painting.' Revised by M. F. Malepeyre; translated by A. A. Fesquet. Pp. xvi., 659. Philadelphia, 1874.

Russell, W. J., and Abney, W. de W., 'Action of Light on Water-Colours' (Blue-Book). Pp. 78. Appendix of 14 diagrams. London, 1888.

Theophilus, or Rugerus, 'An Essay upon Various Arts' (eleventh century). Translated, with Notes, by R. Hendrie. Pp. xxxviii., 447. London, 1847.

CONTENTS

PART III

PIGMENTS

PART IV

METHODS AND RESULTS

THE CHEMISTRY
OF PAINTS AND PAINTING

INTRODUCTION

THE materials employed by 'picture-makers' are now very numerous. Some of the old pigments, and painting-grounds, and methods, have indeed fallen more or less completely into disuse; but, on the other hand, many new products, both natural and artificial, have been added to the resources of the artist, while several new processes of painting have been introduced, or old methods modified. Nowadays it is very seldom that a painter prepares for himself any one of the materials which he uses, generally accepting, without much hesitation and without examination, the paper, the canvas, the paints, the oils, and the varnishes which his colourman supplies, provided they respond, at first sight, to his requirements. True he has abandoned, not without regret, several of the most treacherous compounds by which his immediate predecessors were seduced. 'Pure scarlet' he has given up; he is shy of asphalt; tobacco-juice and Spanish liquorice are no longer regarded as desirable water-colours. He may go so far as to reject chromate of lead, but he still employs the pigment called chrome green, or green cinnabar, for he does not know that the same chromate of lead enters largely into its composition; and he still thinks that madder yellow is a sound paint, because it is called madder, while he rejects the yellow lakes, which are derived from the same source. His linseed oil is neither made from pure linseed, nor cold-drawn; his copal varnish may not have a particle of pure copal in it; but both are taken on trust. I do not expect that artists should become chemists trained to test their materials, but they will place themselves in a position of comparative security by acquiring an elementary knowledge of the origin, the composition, and the characteristics of the various products with which their works are constructed. An architect is expected to recognise the sound or unsound quality of the timber, the stone, the brick, the iron, with which the edifice he designs is constructed: why should the painter take everything on trust? The purchaser of a picture ought not to be distressed by doubts as to its stability. The concentration of the artist's attention on the definitely artistic side of his practice must, of course, be in no wise interfered with, but time may still be found for the acquisition of such knowledge of his materials as shall enable him to discriminate between the good and the bad. He may even try, with great advantage, a few simple experiments—experiments performed in a few minutes with the simplest apparatus, and with the most innocent of reagents. These are the more necessary now that painters no longer buy their raw materials, or make their own paints, and oils, and varnishes, or prepare their own canvases and panels. Before colourmen undertook such work, early in the seventeenth century, painters were eager after receipts, and, there can be no doubt, were ignorant of reasons: there was little exact science underlying their art. Yet it would be unfair to the best colourmen of the present day to assume that they do not endeavour to provide, as far as possible, sound materials. But they do not manufacture all they sell. They are not paper-makers, nor, as a rule, are they manufacturers of oils and varnishes. Many of the pigments they furnish are not of their own make. If, for instance, you inquire the source of the artificial ultramarine you purchase of your colourman, you will find that it has probably been made in a factory wholly devoted to the manufacture of that pigment. The production of this material can indeed be properly carried on only in special establishments thoroughly equipped for a peculiar and difficult work. In reality, this specialization ought to be, and generally is, advantageous, but it renders the position of the colourman somewhat difficult. He has to assume responsibility for the soundness and genuineness of many products of the history and preparation of which he knows little or nothing.

After these introductory observations, I may refer the reader to the table of contents for the plan of the present book, and to the preface for the object with which it has been prepared. I would add, here, only this one remark, that the materials with which a painting is constructed are described in definite order, beginning with the ground, then passing on to the medium and the pigments, not omitting the final varnish, and finally closing with a brief summary of methods of painting, and of the experimental studies by means of which the conclusions given in the earlier portions of the volume have been reached.

PART I

PAINTING-GROUNDS

Chapter I.—Paper, Vellum, Ivory. Chapter II.—Plaster or Intonaco, Stone, etc. Chapter III.—Panel. Chapter IV.—Canvas.

CHAPTER I

PAPER, VELLUM, IVORY

As paper is used as the painting-ground for the vast majority of works executed in water-colours, and as this method of painting offers but slight protection to the pigments employed against hostile influences, it becomes of the greatest importance to ascertain that no unnecessary elements of danger are introduced in the paper itself. We will now proceed to consider briefly the sources and constituents of drawing-paper.

Linen from the common flax (*Linum usitatissimum*) and in the form of white rags is, or should be, used in the manufacture of paper-pulp. Other vegetable fibres might no doubt be successfully employed for this purpose. For instance, the Japanese paper, prepared from the bast-fibres of the paper-mulberry (*Broussonetia papyrifera*), were it made less absorbent by the introduction of a sufficiency of size, would probably become an efficient, strong, and durable substitute for linen-paper; but at present it is the latter kind with which alone water-colourists are practically concerned. Its manufacture in Europe dates as far back as the latter half of the twelfth century.* To the kindness of Mr. W. H.

* Mr. Weale, Keeper of the Art Library in the South Kensington Museum, informs me that the two first paper-mills in France were set going near Ambert, in the valley of the Valeyre, by men who, during

James Weale I am indebted for an opportunity of examining two early specimens, obtained from the 'Registre des Revenues de l'Évêché du Puy.' As one of the sheets contains contemporary entries of the year 1273—the other entries belonging to 1289—these papers are, at least, as early as the years named. Both papers present the creamy hue, the translucency, and the gloss of vellum. One hundred square inches of the earlier specimen weigh 127 grains; of the later, 163. Both are heavily sized with paste made from wheaten starch. The use of starch for sizing paper has been revived of recent years, but animal size or jelly is still extensively employed. Some paper is, indeed, made from felted linen pulp alone without size; but it is blotting or filter paper, and is quite unfitted for water-colour work, for when a wash of pigment is passed over it, the colouring matter and the water partially separate, while the outline of the brush-stroke is not preserved.

Before entering further into the question of what are the essential and what the accidental and unnecessary constituents of paper, I give the summarized results of four analyses, which show the percentage proportions found in good samples:

ANALYSES OF DRAWING-PAPERS

	English, 1876	*Dutch*, 1876	*Whatman*, 1885	*Arnold*
Water	10·9	11·0	7·4	7·4
Size	6·1	4·8	6·3	7·6
Ash	1·1	0·9	1·1	1·5
Fibre	81·9	83·5	85·2	83·5

The four constituents of paper named above may be regarded as essential. The water or hygroscopic moisture

their captivity in the Holy Land, had been forced to work at the manufacture of paper at Damascus. One of these French mills was called 'Damascus,' the other 'Ascalon.' This was previous to the year 1189.

varies inversely with the temperature, and directly with the amount of water-vapour in the air. There is no doubt that if it could be wholly excluded, the larger number of changes which occur in the pigments of a water-colour drawing would be prevented. It is most injuriously active when a framed drawing is exposed to considerable ranges of temperature. Under these conditions the moisture of the paper is first partly turned into vapour, then condensed on the glass, and, lastly, is re-absorbed by the paper, and, for a time, especially by the pigments lying on its surface. This temporary condensation of an excess of moisture upon the coloured surface does much injury before hygroscopic equilibrium is once more re-established Much less harm would accrue were the vapourized water allowed to escape.

The size must be considered next. It may be applied to the pulp or to the sheet, and may consist of gelatine with a little alum, of colophony and soda, of resin, soda, starch, and alum. From good drawing-papers, which are sized in the sheet with animal size, the greater part of the size may be extracted by means of boiling distilled water, the solution being usually neutral, but sometimes faintly acid or faintly alkaline, to test-papers. Gelatine and starch, to the extent of about 5 per cent. of the weight of the paper, are the safest sizing materials.

The ash or mineral matter in paper may be derived from three sources, namely, traces of the original mineral substances taken up by the flax plant from the soil, and still remaining associated with the felted pulp; the mineral matters, such as soda and alum, introduced with the size; and, lastly, the mineral compounds used to whiten, to weight, or to finish the paper, or in bleaching the fibre and as 'antichlors.' In common and adulterated papers the ash greatly exceeds 1 per cent., twelve parts per hundred of paper being no unusual proportion. This 'filling' may contain or consist of the following substances: kaolin or china-clay, silicate of lime or 'pearl-hardening,' chalk or whitening, lead-white, baryta white or 'white dressing,' and artificial gypsum or 'satin-dressing.' Other substances which increase the amount of ash left when a paper is burnt are blue colouring matters, introduced to counteract the natural yellow tint of the pulp. These include: artificial ultramarine, smalt or cobalt blue, and Prussian blue.

What is put down as fibre in the analysis of paper previously given, is a substance, or group of substances, to which the name of cellulose is given by chemists. Cellulose consists of the three elements, carbon, hydrogen, and oxygen; it is, when pure, entirely combustible, leaving no ash.

The most useful chemical tests to be applied to drawing-paper may be thus indicated.

A piece of paper, having a surface of 100 square inches, is weighed in its ordinary hygroscopic condition; it is then heated in the water-oven to the boiling-point of water for one hour, then rolled up into a glass tube, having a sliding glass tube for cover, and again heated to 100° C. for ten minutes. It is allowed to cool, in the double tube arrangement mentioned, in a desiccator, wherein the air is kept dry by means of a water-absorbing substance, such as oil of vitriol. Finally, it is weighed, still in the tubes. The weight of the latter having been previously ascertained, the weight of the dry paper can be found. The difference between this weight and that of the original paper represents the water which it has lost: the loss should not exceed 12 parts in 100.

100 grains of paper should not lose, by repeated extraction with boiling distilled water, more than eight per cent. of the original weight.

100 grains of paper, when burnt to a white ash, should not leave more than 1·5 grain of incombustible residue.

A few hundred grains of paper should be exhausted with warm distilled water: it is best to leave the water and paper in contact for an hour at least. The watery extract is submitted to the following tests: To one half of it a few drops of tincture of dahlia or violet flowers is added: if the purple tint of the tincture changes to blue, alkali is present; if to red, acid; if it is lightened or bleached, then the presence of chlorine, of sulphurous acid, or of thiosulphate of soda is indicated. A sound paper will yield an extract which will exert no action upon a sensitive vegetable colouring matter. The remainder of the warm-water extract is tested for alum or for the chloride of aluminium, by adding to it a drop or two of liquor ammoniæ and then boiling it; a flocculent or gelatinous precipitate of hydrate of alumina indicates the presence of one or other of the above-named compounds.

The roughness or smoothness of the surface of the paper, or cardboard, is not without influence on the permanence of water-colours. The pigments become less intimately associated with the smooth surface of a hot-pressed paper than with a comparatively rough natural surface. The rough surface is, however, liable to wider and more rapid fluctuations in the amount of hygroscopic moisture.

Some apparently sound papers deteriorate in strength and tint on being kept. Such changes may occur even when pure linen rags have been used for the pulp; they may be generally traced to the disintegrating action on the fibre of the chemical bleaching agents employed. The development of rust-spots, when not due to the mount or backing of a drawing, arises from the presence of small particles of metallic iron from the machinery having become embedded in the pulp. These particles appear grey, brown, or black; they may be detected by placing a drop of dilute acetic acid on the suspected spot, allowing it to dry, and then moistening the place with a drop of a freshly-prepa ed solution of yellow prussiate of potash. If the particle be iron a stain of Prussian blue will be produced. However, some dark spots consist of blackened grease, or of tar, or of the paper-fungus (*Myxotrichum chartarum*).

Naturally, there is a small quantity of oil or fat in paper; it varies from 3 to 5 parts in a thousand. The difficulty experienced in immediately wetting a surface of paper caused by the presence of this trace of oil may be overcome by first washing the surface with distilled water to which a drop or two of caustic ammonia has been added. A solution of the natural mixture of alkaline organic salts, known as oxgall, effects the same purpose. The use of borax had better be avoided.

As drawing-papers are sized in the sheet they occasionally show a peculiar defect arising from the irregular distribution of the size. In such cases, when the surface is scraped off, an absorbent layer of imperfectly sized pulp is revealed beneath. When such paper is used for water-colour painting the sinking-in and running of the pigments produce disastrous results; but it is easy to guard against accidents of this sort by previously scraping and colouring a corner of the sheet to be used. The peculiarity is generally owing to the too prolonged and slow drying of the sheets of paper after they have been removed from the warm sizing-bath and pressed. The solution of size is brought to the surfaces from the interior of the sheet, and remains there. Moreover, in very slow drying, the size is apt to decompose with loss of its glutinous character and, possibly, the formation of mildew. A good drawing-paper will indeed have rather more size at the surface than in the interior, this result being secured by a rate of drying which is neither too rapid, nor too slow.

As to vellum, parchment and ivory, little need be said. All three contain the characteristic ingredient ossein, an insoluble nitrogenous organic substance, which by long boiling with water is converted into gelatin: a solution of gelatin constitutes ordinary size. Water-colour paints placed upon any of these materials sink either very slightly, or not at all into their substance—a very few, such as aureolin, strontia-yellow, and madder carmine, stain the superficial layer. The old method of preparing vellum for the reception of water-colours consisted in rubbing the surface with very finely-ground bone-ash, or with pulverized sandarac. Pumice-stone or cuttle-fish, reduced to a minutely divided state by pounding, grinding, and sifting, may be used for this purpose.

Much care is necessary in selecting tinted and coarse-coloured papers for water-colour work. The tints of the former are often obtained by the introduction of fugitive pigments into the pulp; the latter are often made of inferior and mixed fibres, and sometimes contain lead-white and other injurious *fillings*. Mill-board is often made of wood-pulp and straw-pulp: its surface is primed for oil-painting in the same way as canvas.

CHAPTER II

PLASTER, GESSO, STONE, SLATE, ETC.

THE painting-grounds to be considered in this chapter consist mainly of mineral substances. However their constituents may be varied, in accordance with the process to be used in painting upon them, the wall or backing upon which they are spread should fulfil certain conditions. It must be naturally dry, free from soluble saline matters, and not very porous. A damp-proof course above the level of the ground is necessary, and the wall should be well-built, and free from tremors. A double wall well-bonded has been recommended: in this case the air enclosed between its two divisions should not be stagnant.

Before being plastered, the wall, whether its surface be of stone, bricks and mortar, roughened slate, or tiles, must be thoroughly wetted with rain or distilled water. The plaster is applied in two or more coats, the coarsest and thickest first. In the case of a ground for fresco the two ingredients usually employed are (or rather were) pure, slaked lime and clean, sharp silicious sand. The sand must be uniform in grain, white, and free from soluble salts. The slaked lime is so important an ingredient in the majority of plasters, that it is expedient to describe its preparation once for all. Before doing so we may state the relations subsisting between the three compounds known generally as carbonate of lime, or mild lime, burnt lime, and slaked lime. The first of these is neutral and nearly insoluble in pure water, the second and third are alkaline and caustic. When burnt lime unites with water to form slaked lime it becomes slightly soluble in pure water. In chemical language these three compounds are called respectively calcium carbonate, calcium oxide, calcium hydrate (or hydroxide). From the first substance the others are readily obtained. If calcium carbonate, often called carbonate of lime, be heated to a sufficient temperature, it is decomposed, being resolved into carbon dioxide (carbonic acid gas) which escapes, and calcium oxide (lime) which remains: from 100 parts by weight of the carbonate 56 parts of lime, that is, burnt lime, are obtained. Placed in water or exposed to moist air this burnt lime combines with water, 56 parts of it uniting with 18 parts of water to yield 74 parts of slaked lime, calcium hydrate. In the ordinary country atmosphere, which contains no more than 3 measures of carbonic acid gas per 10,000, slaked lime or calcium hydrate loses its combined water, slowly becoming once more the carbonate from which it was originally produced: 74 parts of hydrate lose 18 parts of water and combine with 44 parts of carbonic acid, and yield 100 parts of carbonate. Thus mild lime is formed once more from caustic lime. By this change, if it be effected in the presence of a sufficiency of free water—that is, if the hydrate of lime be in the state of a firm paste—the whole substance becomes a hard crystalline solid, like an opaque marble. Advantage may be taken of this hardening or cementing process to firmly incorporate other substances with the lime. Silicious sand, infusorial earth, pumice, marble powder, and many other mineral substances, may be thus introduced. Such of these materials as are silicious may contain silica in a form which is known as 'soluble silica.' This substance further strengthens the plaster by forming with a part of the lime an insoluble compound called silicate of lime. To return to the preparation for artistic purposes of hydrate of lime. White or black marble, limestone, chalk, or other fairly pure forms of carbonate of lime are first of all *burnt*, and then the quick-lime produced is slaked with clean water. This is done in a *grouting* box, having a sluice 1 or 2 inches from the bottom. Run the thick cream of lime into a tank of slate and keep it, covered loosely, for two months. At the end of this time it will be ready for all the rougher purposes of plastering. For finer work, the grouting operation is to be repeated, and the cream of lime strained through hair-sieves, and preserved in screw-top stoneware jars. Some water will accumulate above the lime-putty, as it may be called, in these jars ; it should be poured off, or drawn off, from time to time. The jars are kept tightly closed to prevent further carbonation of the lime hydrate. This change, if carried beyond a certain point, is undesirable, since the binding and hardening powers of the lime would thereby be lessened seriously, or even vanish altogether by its conversion into mild lime : not more than one-third or at most two-fifths of the lime should be converted into the carbonate. The lime-putty thus prepared may be used for plaster and intonaco with the certainty that it will not give rise to defects in the painting grounds made therewith. Much lime paste of this kind was prepared for the works in fresco in the Houses of Parliament, and was kept in the cellars under that building, where probably some of it still remains. I have made many experiments with samples from that source, and can speak with confidence of its excellent quality.

A good mixture for the first application to the moistened

wall consists of 2 parts (by weight) of clean sharp sand to one of lime-putty. When one or more coats of this mixture have been duly laid and have set, then the surface is ready to receive the final coat or intonaco, the actual painting-ground. Before this is applied, the rougher plaster below must be thoroughly wetted with distilled or lime water. The sand in the intonaco is of finer and more uniform grain than that previously employed; the intonaco itself is only one-eighth of an inch in thickness. All the coats must be laid without having recourse to scraping or 'floating;' the latter operation brings too much lime up to the surface. Considerable practice and manual dexterity are needed in these operations. The work of painting is at once commenced when the intonaco has been laid, no more being spread at one time than the artist can cover in the day. Upon the wet soft plaster the cartoon is laid, and the outlines and other important parts pounced in, transferred, or impressed by an ivory point. Rapidity and firmness of execution, with the distribution of an uniform thickness of pigment, are matters to which special attention must be paid. The chemistry of this method of painting will be discussed in Chapter XXIII.

Many modifications in the preparation, proportions, and materials of fresco painting-grounds have been introduced or suggested from time to time. I have found the following mixture to yield an excellent plaster for this purpose: Three parts of burnt lime in very fine powder are ground up with 2 parts of whitening or prepared chalk; the mixture is grouted, and then strained through hair-sieves; 5 parts of the putty thus obtained are mixed with 5 parts of sifted crushed marble, or with 5 parts of sharp, fine, sifted sand, or with 3 parts of sifted pumice, or with the same quantity of infusorial (silicious) earth; the whole being moistened with a sufficient quantity of lime-water to render working easy. For the undercoats the sand, etc., introduced may be coarser; while a small quantity of the most silky and whitest asbestos, cut with scissors into short uniform lengths, will prove a desirable addition. The asbestos lessens the risk of any lack of continuity in the undercoats.

For fresco-secco the same ground as that required for true fresco may be used, but it is allowed time to dry and harden. So long as it contains any caustic lime this ground is unfitted for work in tempera, as its alkaline nature seriously limits the variety of pigments which may be employed in this method. When carbonation of the lime is complete it may be employed for tempera-painting, the surface being first treated with warm size. Many Greek and Byzantine paintings were, however, executed upon a caustic lime ground, but the pigments employed consisted chiefly of those natural earths which are unaffected by alkalies. In these Greek tempera-grounds slaked lime mixed with chopped straw, flax, or cotton, formed the basis of the plaster. It is scarcely necessary to remark that these vegetable materials are liable to decay and to cause discolouration of the ground.

The ordinary ground for Italian and Spanish tempera-paintings consisted either of whitening and size, or of burnt gypsum stirred well with water so as to lose the power of setting, strained, and mixed with size. Sometimes both whitening and slaked burnt gypsum are found together as constituents of the ground. The ground was laid directly on the panel, or on the cloth which had been previously glued to the wood. Great care was taken by sifting and washing to secure the fineness and purity of the whitening (calcium carbonate) and of the slaked plaster of Paris (calcium sulphate united with two proportions of water).

Various kinds of size were used; one of the best was made partly from parchment, partly from fish-glue. An excess of size will cause the ground to crack; it must never contain such a quantity as to be rendered non-absorbent. All tempera-grounds of gesso were originally absorbent; in course of time they have become more so owing to the decay of the size. Whether they were afterwards to be painted in tempera or oil they were always first sized. This sizing preserved the luminous whiteness of the ground, which was unable to absorb the oil of oil-paints or that present in the egg-yolks employed in tempera. A proof of the existence of this layer of size above the ground proper is obtained in the process of transferring old tempera and oil pictures to canvas, for in such cases we find discolouration of the ground under cracks only where both the size and the paint above it have become fissured.

It will have been gathered from what has been stated in the preceding paragraph that a non-caustic tempera-ground is suitable for work in oils. In the latter case, however, it must be perfectly dry before the painting is commenced. It should be gently warmed and rubbed with a little clean spirits of turpentine before laying on the first coat of oil-paint.

The ground for stereochromy has been modified several times since the first introduction of this method of water-glass painting. Originally it was recommended to use an undercoat containing 2 parts of sharp sand, 2 parts of fine sand, and 1 part of slaked lime in fine powder. Upon this was laid an intonaco of one-tenth to one-eighth of an inch in thickness, made of 3 parts of fine sand and 1 part of slaked lime. The fineness or coarseness of the sand in the intonaco must, however, be regulated by the nature of the surface required by the artist. I see that nearly twenty years ago I recommended* the employment of sifted white marble powder, and of several other substitutes for sand, recommendations which, within the last few years, have been again brought forward by Herr Adolph Keim. Oxide of zinc may be advantageously substituted for a part of the lime in the intonaco, and it may be added to the pigments. Keim recommends the wall to be first coated with a mixture of 1 part of burnt lime (which is to be slaked with distilled water), and 4 parts of a composition consisting of coarse quartz sand, infusorial earth, and powdered marble. The actual painting-ground, which is from one-eighth to a quarter of an inch in thickness, is made of 1 part of slaked lime, and 8 parts of a mixture of the finest quartz sand, marble sand, marble meal, and infusorial earth. Fine asbestos paper, wetted with lime-water, and firmly pressed by rolling into a soft freshly-laid lime and sand plaster, makes an excellent ground for stereochrome painting; but as a single breadth only of this paper can be used, the size of the work that can be executed upon it is somewhat limited. All the precautions as to dryness of the wall and purity of the materials, already noted in the case of fresco-grounds, must be observed in reference to those intended for stereochrome painting.

The ground recommended by the late Mr. Gambier Parry for that modified form of varnish-painting to which he gave the name of 'Spirit-Fresco' is identical with that required for true fresco. All the usual precautions as to the dryness of the backing, and its freedom from soluble salts, must be taken. The plaster must be allowed to dry completely before the operation of saturating it with the medium is commenced; the lime in it should also have become mild—that is, carbonated. (See Chapter XXIII. for tests for

* 'Chemistry of the Fine Arts' in Cassell's 'Technical Educator.'

alkalinity and moisture) Syringing the plaster with distilled water previously charged under pressure with carbonic acid gas, though it delays the drying, hastens the carbonation of the lime materially. To complete the preparation of the ground, it should, when quite dry, be soaked with a mixture of two parts of the medium (Chapter XII.), and three of turpentine. After two days, this treatment must be repeated. A third application may be needed for very porous grounds. Another period of forty-eight hours having elapsed, the surface receives a coat of white paint, made of equal parts of white lead and gilder's whitening, ground up with the medium diluted with one-fourth or one-third its bulk of turpentine. This priming is repeated when the first coat is dry. After three weeks, the painting may be commenced. Stone and terra-cotta, if sufficiently porous, may be primed in the same way as plaster. Under no circumstances should cements containing plaster of Paris be introduced into the grounds used for spirit-fresco.

Slate may be used as the ground for oil-painting; but its freedom from crystals of iron-pyrites, which present a brass-yellow colour, must be first ascertained. The firm adhesion of any priming or other layers of oil-paint which may be applied subsequently to slate may be secured in the following manner. The slate is slowly warmed in a water-oven, and thus becomes quite dry. While still warm, it receives a thin coat of oil-copal varnish, largely diluted with turpentine, or it is rubbed with a bit of hard paraffin-wax. When this is hard, the painting may be carried out as in the ordinary way of using oil-colours; a priming of flake-white ground in oil and mixed with a little copal-varnish and turpentine, may be first applied, if desired. Terra-cotta and stone may be treated in the same way; but, being more absorbent than slate, the process recommended at the close of the preceding paragraph is preferable.

CHAPTER III

PANEL

WOOD, as a backing for the painting-ground of works in tempera and oil, presents some advantages over plaster and canvas. Its chief merit lies, perhaps, in its comparative immunity from mechanical injuries. The wood selected must be hard, that its surface may resist blows and abrasion: and it must not contain much resin, gum, colouring-matter, or other 'extractives,' as they are called, or else discolouration of the painting-ground, or priming, may occur. The Flemings used oak; the Italians white poplar. But oak often proves treacherous, through irregular shrinkage; while poplar is too soft. Italian painters employed, also, the wood of the stone pine. Leonardo da Vinci recommended cypress, pear, and service-tree. Mahogany, which was unknown to the old painters, is now generally employed. Teak and cedar, and also American or black walnut, deserve further trial. Wood grown in poor soils, in temperate climates, and felled in winter, is the best.

The specific gravity of wood varies from 0·3 to 1·3; the lighter kinds contain large volumes of interstitial air. The longitudinal contraction of wood is much less than the transverse; the distribution, form, and number of the cracks in old panel-pictures is often to be traced to this cause.

Wood contains (1) water, (2) ligno-cellulose, (3) extractives,

(4) ash or mineral matter. The water, in thoroughly-seasoned and air-dried wood, generally constitutes about one-eighth part of its weight. The main constituent of wood is the so-called ligno-cellulose, which is present to the extent of from 75 to 85 per cent. It may be resolved into two substances, which, for convenience' sake, are here called cellulose and lignose. The extractives belong to two groups —one, soluble in alcohol and ether, consists chiefly of resins; the other, soluble in cold or hot water, or else in very dilute alkalies, includes tannin, albuminoids, gum, and colouring-matters. The following analyses of three kinds of wood in an air-dried state will convey a fair idea of their constitution in 100 parts:

	Mahogany	*Oak*	*Pine*
Water	12·4	13·1	12·9
Cellulose	49·0	39·5	53·3
Lignose	27·6	34·3	28·2
Ash	1·1	1·2	0·3
Resin	1·0	0·9	1·6
Water-Extract	8·9	11·0	3·7

The preparation of panels for painting requires much time and trouble. The directions given by ancient authorities are numerous, and not always accordant. One author tells us to boil the wood; another says we are to coat it with mastic dissolved in twice-distilled turpentine and mixed with white. Then it is to be treated twice or thrice with spirits of wine, in which some white arsenic or corrosive sublimate has been dissolved; coats of boiled oil, of liquid-varnish* and white, and of verdigris and yellow are subsequently mentioned. Probably the best method of treating the harder woods intended for pictures is, after thorough seasoning, first of all to reduce the panel, by planing and glass-papering both sides equally, to the desired thickness.

* Made by boiling 1 part of sandarac in 3 parts of linseed-oil.

The panel is then soaked in water heated to 50° C., and then steamed. When dry, it receives a wash on both sides of a solution of corrosive sublimate in methylated spirit; it is again dried and seasoned in a warm air-chamber. After these operations, the panel should not require more than a slight rubbing with fine glass-paper, in order to render both surfaces plane. For panels to be used for oil-pictures, a priming is now applied, consisting of white lead, a little copal-varnish, and drying linseed-oil prepared by means of borate or oxalate of manganese. Allow this coat, which is intended to fill up the cavities and pores of the wood, to dry thoroughly, and then apply another coat in the transverse direction; subsequent coats should contain nothing but white lead (or other pigment) and the drying oil. Repeated smoothings of each coat, when hard, with fine pumice-powder are necessary; the last coat may consist of zinc-white and drying-oil. Both sides of the panel should be treated, as far as possible, alike, so that they may be equally loaded, and equally protected; but the pumice-rubbings are, of course, not required for the back of the panel. The object of priming the back is twofold—the prevention of decay, and of the attacks of insects; and the avoidance of that gradual curvature whereby the protected front becomes convex, and the unprotected back concave. This change occurs through the slow loss of water from the back of the panel—a loss which is generally accompanied by a loss of some of the organic constituents of the wood through oxidation. Here it may be mentioned that the original steaming of the panel removes some of the extractives, and coagulates the albuminoids present, which are generally the first cause of decay. This decay is not primarily a chemical and spontaneous one, but is commenced by certain minute organisms, the growth and increase of which is, in part, dependent upon the presence of available albuminoids, but which involves also the destruction of some of the other extractives, and even of the ligno-cellulose itself. The corrosive sublimate employed helps to sterilize the wood, and to prevent the inroads of animal organisms.

In order to avoid the disastrous effects of transverse shrinkage upon compound panels, the old painters glued linen cloth, or vellum, or parchment, or tinfoil to the front surface of the wood, and on this they spread their gesso, or painting-ground. Gesso, made of plaster of Paris and size, or of whitening and size, often lost its cohesion through the decay of the binding material, and in consequence became fragile and powdery; the panel itself decayed, and thus at last the linen or parchment remained as the best preserved element of the composite structure. Were we to avoid gesso and use lead-primed canvas glued to panel, we should really be painting upon canvas backed or protected by wood. Panel is to be recommended for modern work only when a single piece of uniform and well-seasoned wood of sufficient size can be secured. However, an excellent cement for joining panels together was sometimes used with success. It consisted of lime and cheese, both in fine powder, the latter having been grated, and then washed with water. These materials intimately mixed and then ground into a paste with water, yield a tough and adhesive cement which becomes of rocky hardness.

In order to prepare a panel for tempera work, it should be treated in the manner above described, substituting for the priming with oil, white lead and copal-varnish, a mixture consisting partly of parchment-size, partly of fish-glue, and whitening.

It is very probable that some of the hard, fine-grained woods of British India will furnish excellent materials for picture-panels. At present experiments in this direction cannot be regarded as more than tentative and promising.

CHAPTER IV

CANVAS

THE usual, and probably the best fibre for the manufacture of canvas for painting is unbleached flax—that is, linen; hemp and cotton are decidedly inferior. The material is woven in different ways, and with strands of different degrees of fineness, so as to produce cloths of various degrees of thickness and fineness, and having several kinds of texture and surface.

The canvas is first treated with size or a solution of glue; this should be as free from colour as possible. The priming consists of two coats, the first containing whitening and size, the second lead white and linseed-oil. Fuller primings are often given where it is not desired to allow the texture of the canvas to remain evident. Such primings are put on alternately in directions at right angles to one another, and are treated in the same way as the primings of panel. If before the last priming be dry it be dusted with zinc white, or if a final priming of zinc white and drying oil (free from lead) be given, the usual discolouration of the canvas which occurs on keeping it, especially in the dark, will be avoided. But such discolouration can always be removed by leaving in contact with the priming a piece of blotting-paper saturated with a solution of peroxide of hydrogen.

Some painters in oil have employed with success a tempera-priming on their canvases. This priming may be prepared with a mixture of a strong, though elastic, size, with whitening. A good composition of this sort may be made by taking equal weights of fine whitening and of fine plaster of Paris, which has been slaked in and soaked with abundance of clean water, or of the preparation called *satin-finish*, an artificial gypsum, used by paper-makers: the warm size is incorporated with this mixture. When the priming coats are dry the surface is dressed with a layer of pure size, and allowed to harden thoroughly before the picture is begun.

An ordinary primed canvas was examined with the following results. The amount of moisture present was 5·5 per cent. of its weight, the priming 25 per cent., and the dry substance of the size 15. The dry fibre which constituted the remaining part would weigh, therefore, about 44 parts. It was further found, with the same canvas, in a dry heat of 100° C. (212° F.) continued for twenty minutes, that a strip 20 inches long became shorter by a quarter of an inch, changing in colour from a creamy white to a pale buff. After immersion in boiling water for twenty minutes a piece of this canvas 20 inches square was found to have shrunk rather more than 1 inch in one direction, and in the other direction rather more than half an inch. The piece was somewhat crinkled, and had become yellow in patches.

A few remarks as to the bearing of the above observations on some of the phenomena presented by oil-paintings on canvas may be here introduced. The water present in canvas varies with the temperature, and in consequence the dimensions of the canvas vary. As the contraction on drying and the expansion on taking up moisture are not the same in the direction of the warp as in that of the woof, there is an unequal strain upon the layers of paint upon the surface. These may, therefore, become irregularly fissured, and even loosened. The importance of selecting a canvas so woven as to expand nearly equally in both directions is evident, but the maintenance of a uniform temperature, and of a suitable degree of moisture in the atmosphere where pictures are hung, is also obvious. The absorption of moisture by canvas occurs through the back, unless that be also protected by paint. With the moisture deleterious gases may also be absorbed, and these may easily pass through and affect the priming, even the picture. Canvas protected by panel behind, or coated at the back with a layer of white lead which has been ground up with starch paste, escapes this injury in great measure, as the sulphuretted hydrogen, etc., are then intercepted. The colouring-matter of the fibre and size of the canvas may move towards the front and discolour the priming and even the picture. An excess of damp and a high temperature are the chief causes of this movement. When the first priming coat contains size, though it may adhere firmly to the sized canvas, it may not hold the subsequent oil-painting quite so tenaciously. Canvas is liable to accidental injuries from mechanical causes: a double canvas mitigates the evil. The elasticity of the priming may not suffice, when the canvas is rolled up, to prevent cracking. A small addition of a non-drying oil, such as almond or olive oil, to the linseed-oil used in the priming coats, proves useful, but such addition should not exceed 1 part of non-drying oil to 20 of drying oil.

There are two methods of preserving canvas from decay by the application of solutions to the back after the priming has been completed on the front. One of these solutions contains corrosive sublimate (mercuric chloride) dissolved in methylated spirit: a 5-per-cent. solution is sufficiently strong. The other solution is made by dissolving tannin in methylated spirit. One or other of these solutions should be applied once to the back of the canvas by means of a broad stiff varnish-brush. When the spirit has evaporated, the coating of white lead ground in starch paste previously recommended in this chapter may be applied. The two solutions we have named act by coagulating some of the size in the canvas : the tannin turns it into leather. Corrosive sublimate prevents the development of mould or mildew, and is a good preservative against the attack of animal organisms.

It may be observed that the employment of size in the first preparation of canvas constitutes an element of weakness. Many attempts have been made to substitute a less hygroscopic and changeable substance. A rather costly, but elastic and permanent 'filling,' consists of a specially prepared collodion. This is made of the following materials :

Absolute ether - - - -	15 *ounces*
Absolute alcohol - - - -	4 ,,
Pyroxylin - - - - - -	1 *ounce*
Paraffin wax - - - - - -	7 *grains*

The hard paraffin wax is first dissolved in the ether; then the alcohol is added, and lastly the pyroxylin (soluble guncotton). The canvas to be treated is first stretched on a frame, and dried in a hot-air chamber. Then the collodion is rapidly applied to the surface to be primed, several coats being usually required. The oil-primings are added afterwards in the usual way; great care should be taken that each successive coat be dry and hard before the next is laid on. The collodion itself may also be used as a priming. For this purpose it is, of course, thickened and whitened by the addition of some dry pigment, such as Chinese white.

'Willesden canvas' appears to resist the attack of moisture and of animal organisms: how far the copper-compounds and ammonium sulphate which it contains may interfere with its use as a painting-ground has not been determined by experiment. It certainly merits a fair trial. The same remark applies to 'Willesden paper.'

PART II

VEHICLES AND VARNISHES

Chapter V.—Oils. Chapter VI.—Resins, Waxes, Paraffin-Waxes. Chapter VII.—Yolk and White of Egg, Size, Glue. Chapter VIII.—Gums, Glycerin, Honey. Chapter IX.—Water-Glass, Lime- and Baryta-Water. Chapter X.—Solvents and Diluents. Chapter XI.—Siccatives and Dryers. Chapter XII.—Varnishes and Oleo-Resinous Vehicles.

CHAPTER V

OILS

THE common usage of the term 'oil' is wider and less definite than that sanctioned by chemists. We must exclude from the category of true oils petroleum and the liquid paraffins, spirits of turpentine and the volatile essences of plants, the hydrocarbons of coal naphtha, as well as a number of other liquids which present certain superficial resemblances to the oils proper. Fats, however, belong to the same group, their solidity at ordinary temperatures being, so to speak, an accidental rather than an essential difference.

The true oils are often called fixed oils, for they cannot be boiled and distilled without change, thus differing from really volatile liquids. They are *glycerides*—that is, compounds from which glycerin, on the one hand, and fatty acids, on the other, are obtainable. These glycerides are named after the fatty acids which they yield. Thus olein is the glyceride of oleic acid, linolein the glyceride of linoleic acid. In reality three kinds or varieties of glycerides of each fatty acid are possible, but the oils used by painters consist almost entirely of one of these kinds. The formation of one of these glycerides may be expressed in words thus: One molecule of glycerin, reacting with three molecules of a fatty acid, yields one molecule of the glyceride in question and three molecules of water. Conversely, under other conditions, one molecule of a glyceride, reacting with three molecules of water, produces one molecule of glycerin and three molecules of fatty acid. If, in this last reaction, we substitute for the water three molecules of an alkali, such as potash, we obtain glycerin as before ; but, in lieu of the free fatty acid, we find that an alkaline salt of the fatty acid has been formed—such salt is a soap. Alkaline soaps, namely, those of potash, soda, ammonia, are soluble in water, which fatty acids are not. There are, however, other soaps which are insoluble in water, namely, the lime, lead, copper, and many similar metallic salts of fatty acids.

Oils, though insoluble in water, are easily soluble in spirit of turpentine, benzene, chloroform, liquid paraffins, and volatile plant essences ; they are, in fact, miscible in all proportions with these liquids. There are other liquids in which the oils are less soluble, such are alcohol, ether, and glacial acetic acid.

Oils are divisible into two classes, one of which includes those which dry up and harden, forming a kind of elastic varnish, by exposure to the air. The oils of the other class do not harden, but become sticky, and rancid in smell; these oils, however, if submitted to the temperature of boiling water for some time, do in some instances become dry and hard, but the varnish they yield under these circumstances is dark in colour and brittle. The painter's concern is almost exclusively confined to the oils of the first group, generally known as *drying* oils. To the most important of these attention will be directed presently, but the general methods of extracting them first demand a few words of explanation. There are two different processes in use. In one of these, which has been practised widely from very early times, the oil is obtained by pressure; in the other process, invented some fifty years since, the oil is extracted by means of an appropriate solvent. We may dismiss this latter process almost summarily, for the product which it yields, though much greater in quantity, is decidedly inferior to that obtained by pressure. It is less fluid, and contains a larger proportion of solid fats. The solvent commonly employed to dissolve out the oil from oil-yielding materials is carbon bisulphide (CS_2), a compound of carbon and sulphur, which may be prepared cheaply by passing the vapour of sulphur through red-hot charcoal. Of the pressure-process for obtaining fixed oils there are two modifications. In the more usually adopted of these, the oily seed or other material is first heated, and then pressed while still hot; in the other modification the pressure is applied to the cold seed, etc. Heat and pressure give a more abundant yield of oil, but the product is less pure and less well fitted for use in painting. The bulk of the oils of commerce are thus obtained. Cold-pressed oils remain clear in cold weather, are more fluid than hot-pressed oils, and contain a smaller proportion of solid fats and of free fatty acids.

The most important drying oils are those of linseed, poppy-seed, and walnut kernels; others are obtained from niger-seed, sunflower-seed, and hemp-seed. The first place is due to linseed oil.

Linseed oil is obtained from the seed of the common cultivated flax (*Linum usitatissimum*). Linseed varies in size and colour. The usual colours are a purplish-brown and a reddish-brown, but there is a nearly white sort—a mere sport or variety—which may be said to be straw-coloured. It is grown along with the brown variety in some parts of the North-West Provinces of India, particularly in Nagpur, but no pains are taken to keep the strain pure. Through the kind offices of the Director of the Royal

Gardens, Kew, the Government of India were good enough to obtain a specially pure sample of some hundredweights of white Nagpur linseed, and to place it at my disposal. Attempts to grow it for seed in this country and in Belgium failed, but a large quantity of oil was expressed for trial and analysis. Messrs. Bell and Co., of 225, Oxford Street, obtained several gallons of oil by cold-pressure ; many artists have expressed their approval of the product. One advantage of this white seed is the ease with which the purity of a sample may be recognised by the eye, any accompanying weed-seeds differing widely in colour from the white linseed. The skin of the seed is, moreover, thin, the cold-drawn oil is nearly colourless, and the seed is particularly rich in oil, containing no less than 45 per cent. of its weight, although, of course, much less than this proportion is obtainable by cold-pressure. In a hand-press about 25 per cent. was the average yield. Of the common or brown linseed our chief supplies come from Russia and India. The Russian seed is generally finer than the East Indian ; it is, moreover, imported in a less mixed and impure condition. By screening, the greater part of the impurities are or may be removed, but it is sold on a basis of 4 per cent. impurity. The impurities consist of dirt, other oil-seeds, such as mustard, rape, and gold of pleasure, and non-oily weed-seeds. The presence of the last-named, though it reduces the yield, is not otherwise objectionable,* but the same remark does not apply to the foreign oil-seeds. Most of these contain non-drying oils, which mingle with the linseed oil when the sample is pressed and deteriorate its quality.

The percentage of oil in linseed varies between 28 and 45 : by cold-pressure 20 per cent. is the average yield; by

* Occasionally these weed-seeds give up, under pressure, certain matters which deepen the colour of the expressed oil somewhat.

hot-pressure, 27 per cent.; by extraction with carbon disulphide, 33 per cent. The linseed oil in common use by artists is hot-pressed oil, and is very rarely, if ever, obtained from absolutely pure seed. The seed should be kept three months before it is pressed. The expressed oil should be exposed to light in covered glass vessels or tanks, and kept at a temperature of 80° or even 100° F. for some time. It thus loses colour and becomes clear, a slimy deposit being formed. When thus bleached and clarified, the oil should be preserved in corked bottles filled quite full; the longer it is kept the better it becomes for painting, provided the access of air is prevented. The specific gravity of good linseed oil varies very little. At 60° F. (15·6° C.) it is ·935; a bottle which will hold 1,000 grains of water at this temperature will therefore hold but 935 grains of linseed oil. It expands considerably with heat, its specific gravity at 50° C. being ·913 only. One part of linseed oil requires 36 parts of cold absolute alcohol for solution, but only 4 parts of boiling alcohol. It may be purified by solution in boiling alcohol or in petroleum ether. Other methods of purification are generally employed. Amongst these may be named the following: Filtration through felt or carded cotton and charcoal, and then through pyrolusite; agitation with a solution of common salt, followed by washing with water, and drying by a heat of 220° F.; treatment with one four-hundredth part of oil of vitriol, addition of hot water, washing, and drying. Various other processes and reagents have been employed for purifying and bleaching linseed oil. Aqueous solutions of sulphurous acid, green vitriol, potassium permanganate, potassium bichromate, and peroxide of hydrogen may be included in this list. The addition of 1 per cent. of oil of turpentine to the oil, and then passing a mixture of air and steam through it, has also been tried.

Whatever process be adopted, no acid, saline matter, or moisture must be left in the oil. The general and usual result of all the very different kinds of treatment to which linseed oil is subjected, in the above-named and in many other processes, seems to be the more or less complete removal of impurities. The effect on the properties of the purified oil is chiefly seen in its greatly increased rate of absorbing oxygen and consequent hardening.

The chemical composition of linseed oil may now engage our attention. Its ultimate analysis shows it to vary according to the method of extraction adopted, cold-pressed oil containing about 78 per cent. of carbon, 11 per cent. of hydrogen, and 11 per cent. of oxygen; while the hot-pressed oil contains nearly 3 per cent. less carbon, and nearly 3 per cent. more oxygen — linseed oil, extracted by carbon disulphide, is still poorer in carbon, and richer in oxygen. Linseed oil consists chiefly of four glycerides, called, respectively, linolein, linolenin, isolinolenin, and olein. A small, but variable, amount of free fatty acids is also present. The empirical formulæ of the four fatty acids of the above-named glycerides are, respectively:

Linolenic Acid, Isolinolenic	$C_{18}H_{30}O_2$.
Linoleic	$C_{18}H_{32}O_2$.
Oleic	$C_{18}H_{34}O_2$.

Linolein, which is present in linseed oil to the extent of about 20 per cent., is the glyceride of linoleic acid, and has the formula, $(C_{18}H_{31}O)_3,C_3H_5,O_3$; or, as it may be written, $C_3H_5(O,C_{18}H_{31}O)_3$. The relation of this glyceride to glycerin may be seen when the latter body is expressed by the formula, $C_3H_5(OH)_3$. It is probable that the two other main constituents of the oil—linolenin and isolinolenin—are similarly constituted glycerides, and that they closely resemble linolein in physical and chemical properties. When 100 parts of linseed oil are saponified by an alkali, they yield from 9·4 to 10 parts of glycerin.

The most important chemical property of linseed oil, from a painter's standpoint, is its behaviour with oxygen. Under certain circumstances, it absorbs oxygen to the extent of 12 or 13 per cent. of its weight, becoming converted into a mixture of substances for which it is convenient to retain the old name *linoxine*. Linoxine is solid, and not liquid; it is far less soluble than linseed oil in any solvent, and in many liquids it is insoluble. Linoxine is, moreover, denser than the original oil; 100 grains of linseed oil produce about 108 or 109 grains of linoxine. Notwithstanding the greater density of linoxine, when compared with the original oil, its formation is attended by a considerable expansion. In consequence, a layer of raw linseed oil spread upon glass becomes wrinkled during the drying and oxidizing process. During the oxidation of linseed oil, the small quantity of olein it contains remains unoxidized—its presence confers elasticity upon the product. The changes which occur during this oxidation are complex and ill-understood; but there is some formic acid formed, so that the product is sour—carbonic acid gas and water are also produced. There are many ways of bringing about this oxidation. A very common one is to heat the oil to a temperature of at least 100° C., and to blow air through it, or air containing ozone. Many substances favour the absorption of oxygen by linseed oil under the above conditions. Amongst these may be named manganese dioxide, borate, oxalate, or linoleate; red-lead, litharge, or lead acetate; green vitriol or white vitriol, etc. It is better to use one of the manganese compounds, and an excellent result is obtained with the borate of this metal. On the small scale, the operation may be thus carried out: Tie up in a small piece of muslin 20 grains of dry and powdered manganese borate. Suspend the bag in a glass quart flask, into which a pint of linseed oil has been placed, so that the bag is just covered by the oil; lightly plug the mouth of the flask with some carded cotton. Stand the flask in a warm place, where the temperature does not fall below 40° C., nor rise above 100° C. In a fortnight's time, the oil will have become strongly siccative, so that when it is spread in a thin layer on glass, or paper, it will dry up to a tough varnish within twenty-four hours. If the oil and manganese borate be maintained, by means of a water-bath, at a temperature of 100° C., the change will occupy less time, and the product will be just as good; but it is not advisable to *boil* the oil with the borate, although the change may be thus effected in less than an hour. The oxidation may be further hastened by occasionally blowing a little air into the oil through a glass tube kept permanently in the flask. When the rapid-drying quality of the oil has been proved, by experiments made with a drop or two withdrawn for that purpose, the flask is allowed to get cold, and the oil poured into a corked glass bottle, so as to fill it. In the course of the next few weeks, a slight deposit will be formed in the bottle; when this has occurred, the clear oil should be poured off into other bottles, and preserved for use. According to the purpose for which the prepared oil is to be afterwards used, the treatment with the borate must be more or less prolonged; but care should be taken not to carry it so far that the oil becomes ropy or viscous, unless it is intended to make linseed oil varnish. In the subsequent chapters of this book, we shall often refer to this siccative linseed oil as 'manganese oil.' To the above directions for preparing this oil may be added the remark that if the operations be conducted in a strong light, the oil will be bleached, as well as rendered highly siccative. No satisfactory explanation of the action of the manganese borate (and of many other substances used for the same purpose) has been offered. But it seems probable that the absorption of oxygen by the oil is favoured by the removal of certain impurities, and this the borate of manganese may effect.

The increasing specific gravity of the 'manganese oil,' as the process is prolonged, may be used as an indication of the point at which the heating may be discontinued. When the oil has acquired a specific gravity of ·945, it is generally sufficiently siccative for grinding with non-drying pigments, and as an addition to certain varnishes. For these purposes it may even attain a specific gravity of ·96; but when it shows ·99, or ·995, it constitutes a thick varnish, which needs dilution with a suitable solvent. It may be well to remark here that the various processes for rendering linseed oil more rapidly-drying may be regarded as resulting in two actions, partly consecutive, partly simultaneous. The first action, if it could, or did, occur alone, would yield a purified oil *apt* to dry quickly, but very slightly altered in composition; the second action is more profound, and gives rise to a thickened, denser product, in which the drying-process has already commenced. In practice, the first action occurs almost, but not quite, uncomplicated with the second, when linseed oil is warmed with borate of manganese in a vessel to which atmospheric air has very limited access; the second action, which is of necessity associated with the first, takes place when a stream of air is blown through warm linseed oil, even in the absence of manganese borate, but far more quickly in its presence.

The superiority of the highly siccative oils prepared with borate of manganese (or the oxalate) over those in the manufacture of which lead compounds, or white vitriol, are used, is so decided that all description of the older and less satisfactory methods will be omitted. But there are two other ways of rendering linseed oil more siccative, which deserve a passing notice. Into a clear-glass quart-bottle an ounce of distilled water and an ounce of clean iron brads are first placed, and then one pint of raw linseed oil, agitation being avoided. The next day, the bottle, placed in as strong a light as possible, is to be shaken frequently, the shaking being repeated every day, until a drcp of the oil, when tested, shows a sufficient degree of drying character. Finally, the liquid part of the mixture in the bottle is poured into a separating-funnel, and the aqueous part allowed to run away. The oil may require drying and filtration. In another similar process green vitriol is substituted for the metallic iron, the other directions being identical.

The most important property of linseed oil, and some methods for the further development of this property having been discussed, we may now describe the remaining characters of this oil. The cold-pressed oil is very pale straw-coloured, or pale yellow, with occasionally a faint greenish hue; the hot-pressed oil is a darker yellow or brown. The cold-pressed oil, when considerably cooled, remains clear long after the hot-pressed oil has become turbid. The fluidity of the oil is less than that of water in the ratio of 1 : 10. The hot-pressed oil has a much stronger taste and odour than the cold-pressed oil.

The adulteration of linseed oil with other oils may be recognised with more or less precision by means of several different tests. Most of these tests (oil of vitriol test, nitric acid test, etc.) produce reactions in which the oil and the acid acquire varied colours characteristic of different oils. These tests must be applied under exactly similar conditions of temperature, agitation, lapse of time, strength of acid, etc. ; and even then, unless the experimenter is well-versed in the work, the indications obtained are sometimes perplexing and difficult to interpret. The amount of bromine absorbed by a given weight of linseed oil, or better, of the fatty acids obtained from it, affords a valuable test of purity. This amount of bromine is unusually high, much higher than that absorbed in the case of the oils likely to be used as adulterants. But such quantitative determinations can be properly performed only by the skilled chemist. Valenta's acetic acid test is, however, more easily managed. To apply this, take equal volumes, three cubic centimetres of each, of the oil and of glacial acetic acid (specific gravity, 1056·2); mix thoroughly and gradually, heat the mixture until the oil has completely dissolved, or the boiling-point is reached. Immerse a thermometer in the liquid, allow it to cool slowly, and note the temperature at which cloudiness appears. The following temperatures are those at which this turbidity is produced in the case of several different oils :

Name of Oil	*Temperature of Turbidity*
Niger-seed	49° C.
Linseed	57°
Sesame-seed	107°
Almond	110°
Ground-nut	112°
Rape-seed, Mustard-seed, etc.	Not dissolved.

It will be seen from these figures that of these six oils that of linseed is, with one exception, the most soluble; and that the presence of such usual impurities as the oils of sesame, rape and mustard, tends to reduce the solubility, and hence to develop turbidity in the acetic acid sooner—that is, at a higher temperature.

The specific gravity of linseed oil also affords a valuable means of testing its purity. At 15·6° C. (60° F.) it is denser than most other vegetable oils :

Name of Oil	*Spec. Grav.*	*Name of Oil*	*Spec. Grav.*
Linseed - - -	·935	Poppy-seed - -	·926
Gold-of-Pleasure -	·931	Sunflower-seed - -	·925
Hemp-seed - -	·930	*Black Mustard-seed -	·921
*Cotton-seed - -	·930	*Ground-nut - -	·918
Walnut - - -	·929	*Colza-seed - -	·914

The four oils marked with an asterisk are non-drying.

Poppy Oil.—This oil is obtained from the seed of the opium-poppy, *Papaver somniferum.* It is of a very pale straw-colour, often almost colourless, and is nearly free from taste and smell. By filtration through hot animal charcoal it may be completely decolourized. If the fluidity of water be represented by 1,000, that of poppy oil at 15·6° C. is 74. Its specific gravity at the same temperature is ·926. Its chemical composition is near that of linseed oil; it contains the same four glycerides, but in different proportions, for it is mainly made up of linolein and olein. The large quantity present of olein causes poppy oil to be a less rapidly drying oil than linseed. Wolffen, in 1640, stated that poppy oil dries *throughout* in four or five days, while linseed oil forms a pellicle upon the *surface.* Joseph Petitot, writing from Geneva under date January 14, 1644, states that umber is a siccative for poppy oil. Poppy oil was introduced into painting in the beginning of the seventeenth century, after linseed and nut oil. Later on in the same century the Dutch painters acquired greater confidence in this more slowly drying oil, employing it not only in the painting process, but also for grinding their pigments, especially whites, blues, and pale tints.

Nut Oil.—This oil is obtained from the kernels of the common walnut, *Juglans regia.* Leonardo da Vinci directs it to be made from the peeled kernels in order to avoid the chance of darkening its colour, and also causing the subsequent alteration of the tone of the pictures painted with it. The kernels were to be soaked in water first, before being peeled and pressed. The introduction of nut oil into painting followed that of linseed oil, and preceded that of poppy. Cold-pressed nut oil is much paler in colour, and has much less taste and smell than the hot-pressed oil; it also differs in composition much in the same way that cold-pressed differs from hot-pressed linseed oil. The constituent glycerides of nut oil are the same in kind as those of linseed oil, but a larger proportion of linolein is present. Nut oil closely resembles linseed oil in its physical characters: its specific gravity ·929 is intermediate between that of linseed and poppy oil. Besides the three drying oils already described we may name that expressed from niger-seed, *Guizotea oleifera.* It is occasionally employed in grinding artists' colours as a substitute for linseed and poppy oil. Tea-seed and camellia-seed oils, and the oils extracted in Japan from the seeds of *Perilla ocymoides* and from the kernels of *Torreya nucifera*, are not of sufficient importance to demand description.

A few observations may now be offered as to (1) the action of certain pigments on oils; (2) the different amounts of oil needed for grinding with different pigments.

1. *Action of Pigments on Oils.*—The most common action is a physical one, in which the opacity of a pigment is gradually lessened in course of time by the more complete interpenetration of the oil between the particles. Thus yellow ochre and raw sienna, for example, darken in colour because they become more translucent, just as a piece of

oiled cream-laid paper is darker and yellower than the same paper when dry. The light which falls upon it plunges into it more deeply, and on reflection is more highly coloured. In the case of such pigments as we have named, and several others, another cause is at work darkening and modifying the colour: this is the yellowing of the oil itself. And it is the pigments which require the largest proportion of oil for grinding which exhibit in a marked degree the phenomena in question.

A second action between a pigment and the oil with which it has been ground is the peculiar gelatinous or 'livery' condition quickly assumed by some oil-paints. This change is particularly noticeable with the cochineal and madder lakes. I have succeeded in obviating it, by carefully drying the pigments at a temperature just under 100° C., before grinding them with oil, and by substituting for raw linseed oil a mixture of the 'manganese oil,' described in the present chapter, with some poppy oil. Those pigments which dry easily should be ground with more of the latter oil, those which dry with difficulty with more of the former. Sometimes pigments harden quickly in the tube itself; this change is due either to the siccative character of the pigments, or to the introduction of an actual 'dryer,' or to the too copious use of a strongly siccative oil with those pigments which are naturally slow in drying.

The third action between a pigment and the oil with which it has been ground is of a distinctively chemical nature. The most striking example of it known occurs with flake-white. The lead hydrate in normal lead-white saponifies the oil, forming lead-soaps with the fatty acids which it contains, and, at the same time, setting free a small quantity of glycerin. References to this curious action will be found in Chapters XIII. and XXIII.

2. The different amounts of oil required by different pigments may now be considered. As a rule the densest or heaviest pigments require the least oil. A few pigments require an excess of oil in order to protect them from moisture or other injurious agents. Different authorities do not agree at all closely as to the amount of oil needed to make a workable oil-paint from the same pigment. The following list gives the amount required by 100 parts in weight of 19 pigments :

Name of Pigment	*According to M. von Pettenkofer*	*According to Winsor and Newton* (1882)
White Lead	12	15
Zinc White	14	—
Aureolin	—	200
Chrome Yellow	19	32
Yellow Ochre	75	75
Raw Sienna	240	180
Vermilion	25	20
Madder Lake	62	125
Terre Verte	100	70
Viridian	—	75
Prussian Blue	112	75
Cobalt Blue	125	75
Ultramarine (Artificial)	—	37
Raw Umber	—	100
Burnt Umber	—	90
Bitumen	—	126
Brown Madder	—	87
Burnt Sienna	181	195
Bone Black	112	110

The great differences in the above amounts of oil do not cause such serious results in the conduct of the process of oil-painting as might have been expected at first, for they correspond in a measure to the relative bulks of the several pigments. We can use more copal or amber varnish to balance the excess of oil in some pigments, and so secure an uniformity of structure, texture, and rate of drying in the different parts of the work. It is, however, often convenient to remove some of the excess of oil from a pigment before using it, especially with the colours prepared by some makers.* This can be done by leaving the oil-paint on a pad of blotting-paper; but 3-inch cubes of plaster-of-Paris afford a far cleaner and surer method for the absorption of oil. It may be further remarked that the quantities of oil required by some of the pigments in the above table may be reduced by grinding them under greater pressure. Aureolin requires only 80 parts of 'manganese oil' for each 100 of dry pigment, instead of 200 of linseed oil, and then yields a quick-drying and perfectly protected paint. Yellow ochre, raw sienna, and ivory black should be dried at 100° C. just before grinding and then yield workable paints with less oil. The subsidence of vermilion from the oil in which it has been ground may in some measure be prevented by using 'manganese oil' instead of raw linseed oil, and adding to it a small quantity of hard paraffin wax having a melting-point not under 65° C. Some artists find it a good plan to keep their tubes of vermilion and of other heavy pigments in an inverted position—that is, with the cap downwards.

* Dr. H. Stockmaier, of Nürnberg, has found the following percentages of oil in certain oil-paints from different sources which he has analysed:

Flake-White (Roberson and Co.)	16·2
Light Red (Winsor and Newton)	41·9
Burnt Sienna (Dr. Schoenfeld)	59·2
Chinese Ochre (G. B. Moeves)	45

CHAPTER VI

RESINS, WAXES, AND SOLID PARAFFINS

In commercial parlance resins are incorrectly termed gums. The true gums (Chapter VIII.) are either soluble in water or swell up in that liquid, but resins are not acted on by water. The term resin is used throughout the present volume in its proper sense, so that 'copal resin,' 'mastic resin' are spoken of, not 'gum copal,' 'gum mastic.' All the resins used for making vehicles and varnishes are of vegetable origin, and are probably oxidation-products of certain hydrocarbons present in essential oils. Some resins, such as gamboge, contain gum and are called gum-resins; others contain a hydrocarbon (or terpene, see Chapter XI.) or an aromatic acid, and are called balsams; others are true resins, but even these rarely, if ever, consist of a single definite compound, but are mixtures of at least two, often of three, four, or five different bodies. Generally these constituents of true resins differ as to their degree of solubility in various liquids, such as alcohol, ether, spirit of turpentine, benzene, petroleum spirit, and heated fixed oils. They contain carbon, hydrogen, and oxygen, with occasionally a little sulphur, and are usually of an acid character, and are capable of forming soaps, called resinates, with the alkalies. Resins differ much from one another, not only in solubility but also in hardness and in the temperature at which they melt. Those which are least soluble are generally those which are hardest and which require the highest degree of heat to bring them into fusion. Most true resins contain, besides their proper resinous constituents, small quantities of colouring-matter, of water, of crystalline aromatic acids, and of a volatile hydrocarbon or terpene. All these impurities, save the first, may be removed generally with advantage by the following treatment. The powdered resin is thoroughly mixed with a little water and placed in a large glass retort. A current of steam is then passed into the mixture until the terpene and volatile acids present have distilled over. To the contents of the retort carbonate of soda is added (1 part for each 100 of resin). The mixture after agitation is allowed to cool and then filtered through a fine cotton cloth. The purified resin is then washed on the filter with distilled water, then dried in the air and finally in the water-oven: the air-bath and a temperature of 110° to 120° C. may be used for the desiccation of the harder resins.

It might be thought that the subject of resins would be sufficiently discussed from the painter's standpoint by a description of three kinds—amber, copal, mastic. But it will be shown presently that copal and mastic are names given to several distinct substances, and that there are some other resins which cannot be excluded from our view.

Amber is the most familiarly known of all the resins on account of its long use in its natural state for ornamental purposes. Amber beads have not infrequently been found in early British graves; on the Continent these and other ornaments of amber have often been obtained from ancient interments. At Naples I was shown some years ago a very large number of antique fibulæ carved out of this substance: they had just been disinterred from Etruscan tombs. Such amber has often become brittle, especially so far as regards the surface layers; but in other instances the preservation of the properties of this resin has been complete. The chief localities where amber is found are the Prussian shores of the Baltic Sea (particularly between Königsberg and Memel) and the neighbouring plains; it has been found in veins, and is regularly quarried. Some amber, much of it having a dark colour, is found near Catania, Sicily. Near Lemberg (Galicia in Austria) nodules of amber occur in rock. It occurs in several places in Denmark, Sweden, Norway, and France. In the British Museum collection of minerals there is a fine mass from Cambridge. Excellent specimens occur in comparative abundance on the seashore at Southwold in Suffolk, and at several other places on the Suffolk, Norfolk, and Essex coasts. The dark fossil resin found in Birma, often in large masses, is probably not identical with Baltic and English amber. The same observation may be made with respect to the so-called ambers of Travancore in the East Indies, and of the Isle of St. Louis, Senegambia, Africa. In fact, amber, instead of being, as commonly stated, the fossil resin of a single species of tree of Tertiary age, has obviously been derived from no inconsiderable number of different plants. Göppert, so long ago as 1853, satisfied himself that at least eight species of plants besides *Pinites succinifer* have afforded this fossilized resin: he also enumerated 163 species of plants as represented by remains in amber; many others have been since recognised.

Amber has a specific gravity of about 1·07; its hardness is $2\frac{1}{2}$ on the ordinary mineralogical scale. In most of the usual solvents of resins it is either insoluble, or but partially soluble. When heated quickly on a spatula it splits up and then fuses into a viscous liquid, the drops which are formed rebounding as they fall upon a cold surface: this behaviour serves as a distinguishing test between amber and copal. When crushed amber is heated in a retort it fuses at about 280° C. (536° F.), gives off water, succinic acid, marsh gas, a mixture of liquid hydrocarbons (known as oil of amber), and, finally, at a very high temperature, a yellow substance having a wax-like consistence. Sulphuretted hydrogen and other sulphur compounds are also evolved in small quantity, for amber, like several other fossil resins, contains a little sulphur (sometimes ½ a part in 100) in organic combination. Amber breaks with a conchoidal fracture. When fragments of amber are being ground or powdered they emit an aromatic odour. On being rubbed amber becomes negatively electric in a high degree.

It is probable that true amber consists mainly of a single resin (85 to 90 per cent. of the whole) represented by the empirical formula $nC_{10}H_{16}O$. Small quantities of two other resins which are soluble in alcohol and ether, of a liquid hydrocarbon, and of succinic acid are associated with the main constituent, which has received the mineralogical name 'succinite.'

The classical names for amber were ἤλεκτρον, *lyncurium*, *electrum* and *succinum*. In early mediæval times amber was called *vernix*, a term which at first was applied also to sandarac, and later in the fifteenth century to sandarac only, when amber was designated as *glas*, or *glassa*. In modern French amber is distinguished from *ambre gris* as *ambre jaune*. It is the *Bernstein* of the Germans. The word 'amber' is probably derived, through the Spanish, from the Arabic *anbar*, a term applied to ambergris.

Copal is a name given to a number of hard resins which vary not only in their degree of hardness, but also in their degree of solubility: they are the produce of many different species, and even genera of trees, while the origin of several of the kinds still remains unknown. One of the hardest, palest, and best of all is known as Sierra Leone copal, from the port of collection and shipment. It has been identified as the resin produced by a tree, *Copaifera Guibourtiana*, which belongs to the sub-order Cæsalpineæ of the order Leguminosæ. It is probable that the hard West African pebble copal is the resin of the same tree, but it occurs in rolled pebbles with an abraded surface, and is at least semi-fossil: it is collected from the beds of streams. Pebble copal has more colour than Sierra Leone copal, but yields as strong a varnish. The latter resin occurs in irregular rounded lumps or masses, generally varying in size from that of a hazel-nut to that of a walnut. It is hard and elastic. It consists of at least two resins, one of which, present to the extent of thirty-three per cent., is soluble in absolute alcohol and in spirits of turpentine. The other resin constitutes nearly the whole of the remaining part of the copal and becomes soluble in most of the usual solvents, as well as in hot linseed oil, when it has been previously heated to its melting-point or to a temperature of 180° to 221° C. (360° to 430° F.). Another process for rendering this and other kinds of copal soluble is reduction to a fine powder in the presence of water and the subsequent exposure of this powder to the air for several months or even a whole year. The time requisite for this change may be shortened by keeping the powdered copal at a temperature higher than that of the ordinary atmosphere. More will be said as to this and other methods of increasing the solubility of copal in the chapter on Varnishes.

Other species of the genus *Copaifera* yield similar but inferior resins to that produced by *C. Guibourtiana*, but *C. Gorskiana* is the source of Inhambane copal; Benguela copal, Angola copal, and Gaboon copal are other sorts, varying in hue from straw-colour to a dull reddish-orange, produced in all probability by different species of *Copaifera*.

Zanzibar copal is another hard and valuable resin of African origin: it is often called anime. It is produced by another leguminous tree, *Trachylobium Hornemannianum*, which belongs to the same sub-order, Cæsalpineæ, as *Copaifera*. Most of this Zanzibar copal occurs in a fossil or semi-fossil state in the earth near the roots of the trees, or in places where the trees have formerly stood. This fossilized resin is covered when dug up with a semi-opaque, rough, and dull-brown crust; when this powdery coat is removed the remainder of the mass appears of a transparent yellow colour, with a surface covered with small rounded elevations like those on the rind of an orange: this is spoken of as 'goose-skin.' Many of the pieces are flat and tabular, with a thickness of a quarter of an inch or more. The same resin, when occurring on the bark of the living trees of the same species of *Trachylobium*, presents a smooth and glossy surface; it is not so hard as the fossil variety. Zanzibar copal melts at a higher temperature than Sierra Leone copal, and is very hard. In order to render it soluble it may be treated in the same manner as the Sierra Leone copal. Its chemical nature requires further study. The varnish made with Zanzibar copal, though darker in colour, must be regarded as at least equal in strength and durability to that prepared with Sierra Leone copal.

A third resin, sometimes designated as copal, sometimes as anime, is produced by another leguminous tree, *Hymenæa courbaril*, a native of Brazil and other countries of South America. It is rather softer and more soluble than Zanzibar copal. The copal of Madagascar comes from another species of the same genus, *H. verrucosa*. A Mexican copal is probably the resin of an allied species. The resin from *H. courbaril* is generally known as West Indian copal: fine specimens have been received from Demerara.

The Bungo tree of Sierra Leone, *Daniellia thurifera*, affords a resin of inferior quality. It is probable that the same leguminous tree is the source of some of the Niger and Sudan copals.

A rather hard resin of comparatively recent introduction is Kauri or Cowdi copal, produced by the Cowdi pine of New Zealand, *Dammara australis*. This is a coniferous tree belonging to the tribe Araucarieæ. The largest masses, some of them occasionally over 100 pounds in weight, are found in the earth in many places far from those in which the trees now grow. Kauri resin usually becomes more transparent and yellower by keeping. It is generally somewhat whitish, or streaked with opaque bands, when first found. It is cleaned and scraped and then sorted into several qualities. Its use has greatly extended of recent years. It is now largely employed as the basis of most of the so-called copal varnishes, on account of its abundance, its low price, and its easy manipulation. But the varnish which it yields is inferior in hardness, toughness, and durability to that made from Sierra Leone copal or Zanzibar copal.

Kauri resin is sometimes spoken of as dammar, but this name properly belongs to the resins produced by other trees, not by *Dammara australis*. White or Singapore dammar is the resin of *Dammara orientalis*. It is soft, and may be scratched even by mica. 'Sal dammar' is produced by *Shorea robusta*, the sal tree, widely distributed in India. This resin, though soft, yields a good flexible paper varnish. The tree belongs to the Dipterocarpeæ. *Vateria indica*, another Dipterocarp, yields piney resin or white dammar: a similar resin is produced by another species, *V. acuminata*, a Ceylon tree. Several kinds of *Hopea* (*H. micrantha*, *H. odorata*, etc.), which belong to the same natural order, yield pale, transparent resins which are a trifle harder than that of the sal tree. Black dammar or Tinnevelly resin is produced by *Canarium strictum ;* it is of very inferior quality. This tree belongs to the Burseraceæ : several kinds of elemi resin are also furnished by plants belonging to the same natural order. These elemis are soft, sticky resins, occasionally employed in varnishes to prevent them from becoming brittle and cracking. They contain essential oils and other aromatic bodies, and vary very much in composition and properties, although they resemble one another in their solubility in boiling alcohol and in their easy alterability. They are unsatisfactory resins.

The resin first known as sandarac was probably juniper resin, although the name was also applied to amber. It is spoken of by the older authorities on painting as having a red colour. Its hue is a dull reddish orange, and it yields a dark-brown varnish when dissolved by the aid of heat in a drying oil. The effect of this varnish in imparting an agreeable warm tone to pictures painted in tempera is very evident, when the cold aspect of an old Italian unvarnished tempera picture is compared with the glowing colour of a painting which still retains its original sandarac varnish. The resin now called sandarac is produced by another coniferous plant (*Callitris quadrivalvis*), a native of Algiers. It is a pale yellow resin, when fresh resembling mastic in colour, but becoming yellower with age. It is soft and brittle. When finely powdered and sifted it forms one of the kinds of pounce used in preparing the surface of parchment and vellum for writing and illuminating. It melts at 130° C. (266° F.), exhaling an aromatic odour. It partly dissolves in alcohol, and is wholly soluble in oil of spike and in several terpenes ; acetone also dissolves it.

There is one more resin which requires mention. This

is mastic. The best and most important sort of mastic is produced by a small tree (*Pistacia Lentiscus*), belonging to the cashew-nut order or Anacardiaceæ. This tree occurs in Scio and other islands of the Greek Archipelago. Mastic exudes in the form of tears from incisions made in the bark. It occurs in small pea-like masses, and presents when fresh a very pale straw colour. It is so fragile that it may be crushed to powder between the fingers. It has an aromatic odour, and dissolves completely in boiling alcohol and in spirits of turpentine. Its melting-point is low. It contains, besides its resinous constituents, a small quantity of a volatile essential oil (a terpene) and of moisture. It yields a tender but glossy varnish, largely employed for the final protection of pictures in oil. This varnish yellows with age, and becomes fragile and fissured.

Resins, sometimes called mastics, are produced by other trees of the same genus. These resins, which are of no value for artistic purposes, are :—

Indian mastic from *Pistacia cabulica.*
Bombay mastic from *P. Khinjak.*
Pistachio mastic from *P. Terebinthus.*

Wax.—The true waxes, unlike the oils described in Chapter V., are not glycerides, and do not therefore yield glycerin when they are saponified—that is, turned into soaps by the action of alkalies. Ordinary beeswax is the best known, and probably the most important of all the different kinds; but very few experiments have been made as to the utilization of exotic and vegetable waxes in the processes of painting. Crude beeswax requires purification and bleaching in order to fit it for artistic use. The first operation consists in melting the wax at nearly the lowest temperature possible, and then pouring it in a slender stream into a cold saturated solution of alum, agitating the latter all the time. The granulated wax thus prepared may be bleached by exposure for several days on linen cloths to the action of the sunlight and dew; or it may be treated with dilute chromic acid solution, or with hydrogen peroxide. All these processes succeed better when the wax is in the form of thin sheets or ribbons. The bleached wax, after thorough washing and drying, is to be re-melted. Its hardness is increased and its melting-point raised by the above treatment.

Bleached beeswax melts at 62° or 64° C. (144° or 147° F.). It consists of four distinct substances, not present in all samples in the same proportions. By boiling wax with strong alcohol a substance called myricin (myricyl palmitate) is left undissolved. The dissolved portion is the larger; the bulk of it, which crystallizes out as the alcohol cools, was formerly called cerin. It is a mixture of two fatty acids. The cold alcohol still retains a small quantity of a fourth substance.

Beeswax, by long-continued exposure to atmospheric influences, disintegrates and partially perishes by oxidation. It is a constituent of Gambier-Parry's spirit-fresco medium, into which it is introduced in order to impart a matt appearance to the painting. Excellent examples of the use of melted wax as a binding material for pigments may be seen in the National Gallery and the South Kensington Museum. They are *encaustic* portraits, executed probably in the second and third centuries of our era, and were discovered by Mr. W. M. Flinders Petrie, in the Hawara Cemetery, Fayum, Egypt. The pigments were mixed with wax and laid on in the melted state. The wax having become disintegrated in the course of centuries has been re-melted, some fresh wax having been added in several instances.

Wax is abundantly distributed in the vegetable world; its production is, in many cases, stimulated by the attacks of insects. Thus, Chinese wax is produced by the puncture of Coccus Pela living on *Ligustrum lucidum* and *Fraxinus chinensis*. Chinese wax, which melts at 82° C. (180° F.), consists almost entirely of cerotyl cerotate. Brazilian or Carnaüba wax occurs naturally in thin films on the leaves of a palm (*Copernicia cerifera*); it melts at 84° C. (183° F.). Japanese or Ibota wax is probably produced by the attacks of a coccus on *Ligustrum Ibota*; it melts at 42° C. (108° F.).

Paraffin wax, hard paraffin, and solid paraffin, are names given to certain mixtures of hydrocarbons occurring in native petroleum and in the 'mineral wax' called ozokerite, and also in the tars produced by the destructive distillation of wood, peat, lignite, bituminous shales, and coals. The liquid hydrocarbons which accompany the paraffin wax are described so far as necessary in Chapter XI. under the head of *Solvents*.

Paraffin wax contains no oxygen, and is a mixture of several of the least alterable of all organic compounds; very few chemical reagents have any action at all upon it. On this account it presents for artistic purposes a marked superiority over beeswax or any vegetable wax. Of the hydrocarbons occurring in large quantity in paraffin wax the best known are those to which the chemical formulæ $C_{22}H_{46}$, $C_{24}H_{50}$, $C_{26}H_{52}$, $C_{27}H_{56}$, $C_{28}H_{58}$, and $C_{30}H_{60}$ belong. The melting-point of paraffin wax oscillates within wide limits, say, from 30° to 80° C. The higher the melting-point the harder, the heavier, and the less crystalline is the material. For artistic purposes, hardness and the absence of a tendency to separate from solution in the form of large crystals are desirable properties. Unfortunately the hardest paraffin waxes of high melting-point are much less soluble in oils, terpenes, and varnishes than the softer varieties, and thus their usefulness is limited; they are also somewhat yellowish in hue. I have, however, found that a pure paraffin wax from the Bathgate shale, having the melting-point of 65·5° C. (150° F.) answers every purpose. It is sufficiently hard and but indistinctly crystalline, and yet may be dissolved in fair abundance by the usual solvents. It is convenient to preserve it for use in the form of small flattened globules, which are easily prepared by melting the substance and pouring it drop by drop on to the surface of a large sheet of glass previously moistened by breathing upon it. When these drops are shaken in a bottle they rattle like small pebbles, and do not mark the glass; when the softer solid paraffins are thus treated, they fall with a thud, and leave streaks and spots upon the interior surface of the vessel. This difference of deportment affords a ready means of distinguishing between a paraffin wax suitable for artistic uses and one which had better be rejected.

The manufacture or isolation of hard paraffin and its purification are not described here. The processes employed—distillation, treatment with oil of vitriol, fractional crystallization from solvents, etc.—involve the use of complex apparatus. It may, however, be here stated that commercial hard paraffins vary somewhat in purity. Those obtained from mineral wax or ozokerite are nearly free from oxygen compounds; while those derived from the products of the destructive distillation of shales, coal, etc., sometimes contain as much as 3 per cent. of oxygen, indicating the presence of other bodies besides hydrocarbons. Some of these bodies are of an acid nature; these may be separated by repeatedly boiling the commercial paraffins in question with a 5 per cent. solution of caustic potash. The following table shows the relations subsisting between the melting-point and the specific gravity (at 20° C) of six different samples of hard paraffin from ozokerite:

No. of Sample	*Melting-point*	*Specific Gravity*	*No. of Sample*	*Melting-point*	*Specific Gravity*
1	56° C.	0·912	4	72° C.	0·935
2	61°	0·922	5	76°	0·939
3	67°	0·927	6	82°	0·943

Hard paraffin wax may be used in the preparation of painting mediums as a substitute for beeswax; for preventing the separation of heavy pigments, such as vermilion, from the oil in which they are ground; and for the preparation of certain painting-grounds.

CHAPTER VII

YOLK AND WHITE OF EGG; SIZE; GLUE

THE materials described in the present chapter owe their peculiar properties—at least, in great measure—to the presence of chemical compounds which contain the element nitrogen. Now, this element is not a constituent of any of the artists' materials already described, nor, indeed, of any others, except a few pigments, such as aureolin, Prussian blue, and indigo. The presence of nitrogen in an organic compound is very often accompanied by a measure of instability, or proneness to change; the nitrogenous constituents of eggs, and of size, afford illustrative examples. Another source of weakness in the composition of the nitrogenous constituents, both of the white and of the yolk of eggs, lies in the presence of another element—namely, sulphur. Part of this sulphur readily leaves the original substance, yielding simpler compounds, such as sulphuretted hydrogen, and ammonium sulphide, which possess the objectionable property of discolouring many of the metallic pigments used by artists. On the other hand, all these nitrogenous bodies are susceptible of coagulation, whereby they become insoluble, and very much less prone to change. Indeed, the majority of them may be turned into a substance which is virtually leather, a material which resists decay in the most marked manner. This tanning operation may be readily effected by treating the substances in question with a solution containing tannin, the active ingredient of oak-bark, sumach, nut-galls, etc.

We will first consider the composition of the yolk and white of ordinary hens' eggs. The percentage proportions are, on the average:

	Yolk	*White*
Water	51·5	84·8
Albumen, etc.	15·0	12·0
Fat or Oil	30·0	0·5
Mineral Matter	1·4	1·2
Other Substances	2·1	1·5

The white, it will be seen, is characterized by the presence of 12 parts per hundred of albumen, which is in solution in the ropy liquid. When this solution is heated to a temperature considerably below that of boiling water, the albumen becomes insoluble, and is said to be coagulated; it is not capable of being again dissolved. Solutions of tannin, corrosive sublimate, and many other compounds, inorganic and organic, produce a similar effect. But egg-white is not a pure solution of albumen. For all practical purposes in the arts, it may be sufficiently freed from extraneous matters in the following manner: The necessary number of 'whites' are mixed in a wide-mouth stoppered bottle, with twice their bulk of water, and shaken up thoroughly; then a slip of yellow turmeric-paper is dropped into the mixture. Drop by drop weak acetic acid is poured in, until the reddened turmeric-paper has *just*, or *nearly*, regained its original yellow hue. In this way the alkaline reaction of the liquid is almost neutralized, and it becomes thinner. After further agitation, the mixture is poured upon a piece of well-washed muslin in a funnel. The clear liquid which drops through has been freed from membranes, etc., and contains nearly 4 per cent. of albumen. It may be concentrated by cautious evaporation at a temperature not exceeding 50° C. The albumen which it contains is a very complex substance, containing, besides carbon, hydrogen, nitrogen, and oxygen, about 1·6 per cent. of sulphur. A solution of albumen spread upon glass, and allowed to dry slowly at the ordinary temperature, leaves a residue of albumen in the form of a nearly-transparent film. This, when quite dry, is brittle, and easily cracks. If, before it be quite dry, it be heated to 70° or 75° C., it cannot be again dissolved by water, having been converted into the insoluble form. In this condition it is much less prone to change. It will now be seen how powdered pigments, if ground up with albumen solution and then used in painting, may be made to cohere, and also to adhere to the painting-ground of cloth, paper, or plaster, on which they have been spread. And afterwards, by simply heating the work sufficiently, the whole coloured layer may be rendered insoluble and irremovable by water. Advantage may also be taken of the action of tannin on albumen to secure the same result—the coagulation of the albumen. We may coat a piece of fine linen cloth with albumen-solution, and before it is quite dry we may paint upon it with pigments which have been previously ground up with a weak solution of tannin. If the work be carefully done, the colours will, when dry, be found to have been fixed by the reaction between the tannin and the albumen. If, however, the pigments be laid on somewhat thickly, it may be found necessary to give the whole surface a final coat of albumen-solution. We have dwelt at some length upon this employment of tannin, or of heat, to secure the coagulation of albumen, because it serves to illustrate the way in which paintings, executed with egg-yolk, or size, as a medium, may be fixed. For, as we shall now proceed to show, egg-yolk and size possess many characters in common with albumen-solution.

But the yolk of an egg contains other substances besides albumen. First of all, the albumen present is accompanied by another similar compound called vitellin, which closely resembles it in composition and properties, and which, for our present purpose, we need not further describe. Of albumen and vitellin, taken together, egg-yolk contains, as we have seen, not less than 14 or 15 per cent. But egg-yolk is something more than a solution of these two similar bodies. It is, in fact, an oily *emulsion*, in which innumerable minute globules of a thick, fatty oil are suspended in an albuminous solution. And, moreover, the amount of this oil is large; for there is 30 per cent. of it, as against 15 per cent. of albumen and vitellin taken together. Hence it happens that egg-yolk, the usual vehicle for pigments in the best kind of tempera-painting, must be regarded as essentially an oil-medium. As it dries, the oil hardens, and remains intimately commingled with the albuminous substances left behind on the evaporation of the water present. These albuminous substances coagulate and become insoluble in the lapse of time—a change greatly accelerated by the old practice of exposing the finished tempera picture to sunshine previous to varnishing it.

Size and glue may be considered together. They consist of two distinct yet similar compounds, known respectively as gelatin and chondrin. These bodies consist of carbon, hydrogen, nitrogen, and oxygen; and, when pure, they contain no sulphur. They are soluble in hot water, yet are coagulable by tannin and by some other compounds, organic and inorganic. Chondrin is thrown down from its solution by alum, and, indeed, by several compounds which do not precipitate gelatin. The latter body is obtained from skin, tendons, and bones. These organized structures contain a substance called ossein, or collagen, which, under the influence of boiling water, dissolves, becoming changed into gelatin. This conversion occurs more quickly when the process is performed under a pressure somewhat greater than that of the atmosphere, and, therefore, at a temperature rather higher than 100° C. In this way the transformation of the organic tissue of ivory, bone, vellum, parchment fish-bladder, etc., into gelatin may be readily effected. The purity of the product depends, in part, upon the care with which the raw materials have been selected and cleansed, in part upon the temperature and the duration of the extraction. If the temperature be too high, or the boiling be much prolonged, the gelatin produced is transformed partially into a substance which does not gelatinize when its aqueous solution is cooled. Chondrin is obtained from cartilage, which consists mainly of cartilagin, or chondrigen, by the same process which changes collagen into gelatin. A hot solution of chondrin gelatinizes on cooling just like one of gelatin; but it does not yield, with the same amount of substance, so firm a jelly. Size, glue, and commercial gelatin, consist of mixtures of gelatin, chondrin, and the non-gelatinizing substances produced by the long-boiling or the over-heating of their solutions. Isinglass, vellum, and ivory-dust yield a size which contains nothing but gelatin and a little mineral matter; the darker and stickier kinds of glue contain many impurities, having been made from very varied materials, such as ox-hoofs, horseflesh, old leather, etc.; they often contain sulphuric acid.

In selecting a size for artistic usage, the special purpose in view will indicate whether an insoluble (in cold water) and strongly-gelatinizing, or a partially soluble and very adhesive one should be selected. The former is less liable to crack when dry than the latter. A few experiments, with cold water and then with hot, will soon reveal the peculiarities of the samples submitted to examination. As caustic lime, caustic soda, chloride of lime, sulphurous acid, and certain mineral acids, are frequently employed in the manufacture of size, glue, and gelatin, it is absolutely necessary to ascertain, before using these materials in any process of painting, their freedom from free acids, free alkalies, or bleaching agents. A hot-water solution of the material must not redden blue litmus-paper, nor bleach dahlia-paper, nor embrown turmeric-paper.

Glue and size may sometimes be purified and improved by cutting up the solid or gelatinous mass into small pieces, soaking them in distilled water for a few hours, and then pouring off the liquid before dissolving them.

The temporary preservation from putrefaction of the solutions of the substances described in the present chapter, may be effected in several ways. A lump of camphor, or a few drops of eugenol (from oil of cloves), is generally sufficient. I have preserved the egg-yolk medium for tempera-work for many days in an agreeable condition for use by the following plan: A saturated solution of eugenol in five per cent. acetic acid is first made, then this is added, drop by drop, with constant agitation, to the required number of yolks in a wide-mouth bottle, the point at which to stop further addition being learnt by the change of colour of a slip of turmeric-paper. When this paper just regains its original yellow colour, which was turned brownish-red by the yolks, no more acetic acid is wanted. Any water needed for thinning the medium may now be added, together with a lump of camphor, which will remain floating on the surface.

CHAPTER VIII

GUM, STARCH, DEXTRIN, HONEY, AND GLYCERIN

THE term gum is properly applied to a number of non-crystalline, structureless substances, of vegetable origin. They consist essentially of so-called hydrates of carbon, and are either soluble in cold water, or swell up when left therein for some time. The only gum of any importance in painting is gum-arabic. This name is not, however, exclusively applied to one variety only; it is given to the gums which exude from several species of *Acacia*. For instance, *Acacia arabica* furnishes the Morocco, Mogador, Brown Barbary, and East Indian gums of commerce. But it should be noted that, although *A. arabica* is a native of India, and is grown to some extent in many parts of that empire, the gum it yields is rarely, if ever, exported thence, the so-called East Indian gum-arabic being really taken from Red Sea ports to Bombay, and thence re-shipped to Europe. *Acacia arabica*, however, does not furnish a strong and durable gum, and it is from another species, *A. Senegal*, that we obtain the gum employed as a binding material for water-colours. This gum is known commercially as Kordofan, picked Turkey, white Sennaar, and Senegal gum. The tree which yields it is a native of Senegal and the Sudan; it grows to a height of twenty feet. The supplies which come from Kordofan are of the finest quality, but all the grades of gum from *A. Senegal* are superior to the produce of *A. arabica* in their greater dryness, density, and adhesiveness, as well as in the smaller amount of mineral matter which they contain. It may be added in this place that, according to some authorities, a part of the gum Senegal of commerce is produced by other species of *Acacia* besides *A. Senegal*, such as *A. Adansonii*, *A. albida*, *A. dealbata*, *A. nilotica*, *A. Verek*, etc., and even from species of *Kaya*, *Spondias*, and *Sterculia*.

Gum is essentially a mixture of the salts of an acid called arabic acid. The salts are those of the three bases—potash, lime, and magnesia; water is also present. It is probable that, in all varieties, even of the finest gum Senegal, other organic acids, besides arabic acid, are present. An analysis of a fine specimen of picked 'Turkey gum' gave 15 per cent. of water, and 2·8 per cent. of ash, leaving 82·2 per cent. for the arabic and other allied acids and organic matters. The arabic acid is generally expressed by the formula $C_{12}H_{22}O_{11}$, but the experiments of O'Sullivan indicate a much more complex composition ($C_{89}H_{142}O_{74}$).

Gum from *Acacia Senegal*, the only sort which should be employed in painting, should be nearly free from colour, and should dissolve in cold water without leaving an appreciable residue. Its watery solution should be clear, and should give no colour with tincture of iodine, but an abundant precipitate with ammonium oxalate solution. If iodine produce a purplish colour, adulteration with dextrin is indicated; the white precipitate thrown down by the oxalate shows the presence of calcium, a constant constituent of the genuine gum. I have found that the samples of gum sold to me as gum Senegal were of a more pronounced yellowish colour than those bought as gum-arabic and best Turkey: the lumps varied more in size, often contained air-bubbles, and were less fissured. The adhesiveness and toughness of these samples, moreover, compared favourably with these properties as exhibited by the finest and whitest 'Turkey gum' obtainable.

For the preparation of water-colours, and for occasional use in the operations of painting, it is convenient to have at hand a standard solution of gum. This may be prepared by dissolving 1 ounce of the selected gum reduced to fine powder in 2 measured ounces of boiling distilled water. The powdered gum should be very slowly added, with constant stirring, to the boiling water. When the whole is dissolved, the liquid is allowed to stand for at least a day; then it is decanted from any sediment that may have been deposited into a wide mouth bottle without cork or stopper, but covered with a glass cap. It is well to allow a lump of camphor to float in it, or to add to it a couple of drops of eugenol, the active antiseptic constituent of oil of cloves.

Gum tragacanth is produced by certain leguminous shrubs belonging to the genus *Astragalus*. Amongst these may be named: *A. gummifer*, *A. eriostylus*, *A. brachycalyx*, and *A. adscendens*. It is only partially soluble in cold water. It consists of arabin and bassorin—that is, of arabates of the same compounds which constitute the bulk of gum-arabic to the extent of half its weight, along with meta-arabates, which latter probably have the same composition as the corresponding arabates, but are insoluble in cold water: there is also some starch in gum tragacanth. It contains from 12 to 15 per cent. of water, and leaves 2 to 3 per cent. of ash when burnt. A mucilaginous medium made with gum tragacanth may be used for painting on linen: it is not very easy to prepare so as to be of uniform consistency. A fairly good plan is to place the finely-powdered tragacanth in a bottle, and to add enough spirit of wine to moisten it:

then add the required amount of water and shake the mixture gently at intervals. Water containing no more than 3 or 4 per cent. of the gum constitutes a moderately thick mucilage.

Other gums are of small importance. They commonly contain much bassorin (probably identical with the meta-arabates), and but little arabin (or arabates). The Australian wattle gums from several species of *Acacia* are perhaps thus constituted; but if this be the case, the bassorin present in them seems merely to swell up without dissolving even when boiled for a long time with water, and so differs from the bassorin of tragacanth. Cape gum is produced by *Acacia horrida: A. stenocarpa* yields Suakin gum. So far as I know, no use has at present been made, in the fine arts, of these gums.

Starch comes next in our list. This important food-substance occurs in commerce in a condition so nearly pure that there is no need to describe its character. For the limited uses to which it is put in artistic practice the uncoloured or white starch should be selected. The starch from rice, wheat, maize or potatoes may be employed indifferently. Arrowroot may also be used. The preparation of starch-paste does, however, require some care. The best plan is to thoroughly agitate 50 grams of the dry powdered starch with enough *cold* water to produce a liquid of creamy consistence, and then to pour this mixture slowly into a vessel in which about 300 cubic centimetres of distilled water is kept in steady ebullition. All but 2 per cent. of the starch will dissolve into a nearly transparent homogeneous paste: the quantity of starch must be reduced if a thinner liquid be required.

Starch contains carbon, hydrogen and oxygen only, and is a carbohydrate having the empirical formula $nC_6H_{10}O_5$. It is a stable compound. Commercial starch always contains some water, generally from 12 to 18 per cent.

Dextrin, or British gum, as met with in commerce, is prepared from starch in one or other of several different ways, and is a variable mixture of at least three varieties of true dextrin, soluble or modified starch, starch, a sugar called maltose, and certain minor ingredients and impurities. It will suffice for the purpose now in view, if we select a commercial dextrin, free from acidity, dissolving nearly completely in cold water, and then yielding a solution which, even when strong, has only a light yellowish or brownish colour. When a filtered cold-water solution of commercial dextrin is allowed to evaporate on a glass plate, and the residue becomes air-dry, the film of dextrin left differs from one of true gum by being less friable. A solution of dextrin is, however, far less adhesive than one of true gum of the same strength.

Honey now claims our attention. It is a common ingredient in moist water-colours, and was often employed in size-painting. It is used to counteract the brittleness of gum or of size when dry, or, by its absorption and retention of water, to keep a paint moist. Honey consists of nearly equal quantities of two sugars known as dextrose and lævulose, a little sucrose or common sugar, small quantities of non-saccharine compounds, and about 20 per cent. of water. As the useful properties of honey depend entirely upon its lævulose, a solution of this sugar should be employed instead of the raw honey: this may be easily prepared in the following way: Pure pale honey, kept until it has become crystalline and semi-solid from the separation of dextrose, is mixed gradually with four times its bulk of proof spirit, and thoroughly shaken at intervals for a few hours. The pale yellow alcoholic solution is then filtered: the filtrate is a solution of lævulose, accompanied by small quantities of the other sugars of honey and of harmless impurities, and for some artistic purposes is at once available. Should it be desired to obtain a more concentrated solution of this substance, the liquid may be evaporated to the desired consistency in a porcelain basin, or it may be submitted to distillation in a retort. The aqueous solution of lævulose may be decolourized by filtration through warm animal charcoal.

Lævulose, when free from water, forms a glassy solid; but it is usually obtained as a thick syrup. Although this sugar is capable of assuming the crystalline form, it never does so under ordinary conditions. It has a strong attraction for moisture; on this property its usefulness as a constituent of certain paints depends.

Glycerin was discovered in 1779 by Scheele as a by-product in the preparation of lead-plaster; for a long time the comparatively small quantity of glycerin met with in commerce was obtained in this way. It is now prepared from oils and fats by distilling them in a current of superheated steam; sometimes by first saponifying them with alkalies, or decomposing them with sulphuric acid, and then submitting them to this distillation-treatment.

Glycerin generally occurs as a thick syrup with a sweet taste: when pure, it may be obtained in deliquescent crystals. Its empirical formula is $C_3H_8O_3$. It is a strongly hygroscopic or water-attracting substance, the pure water-free glycerin being capable of absorbing more than one-third its weight of water from the air. Commercial glycerin always contains water: the specific gravity of the liquid affords a rough method of estimating the amount. For pure glycerin at 15·6° C. has the specific gravity 1·264, while that which contains 20 per cent. of water is reduced to

1·21; with 30 per cent. it is 1·183, and with 40 per cent. 1·156. The presence of sugar, a not uncommon adulterant, may be recognised by the turbidity caused by mixing the glycerin, after evaporation to remove water, with chloroform. Glycerin containing lead darkens when sulphuretted hydrogen water is added to it, while the presence of acids may be recognised by blue litmus-paper, which is not reddened by pure glycerin.

The water-attracting property of glycerin induced me to use it as a substitute for honey in preparing moist water-colours so long ago as 1856, when I recommended its employment for this purpose to the late Mr. Winsor. Even in cake-colours a trace of glycerin may be introduced with advantage, as it renders them less friable and more easily rubbed down with water. It prevents size, glue, and white of egg from becoming brittle on drying, and on this account may be used in the preparation of linen, canvas, etc., as painting-grounds. Care must, however, be taken in every case not to add more glycerin than is necessary to effect the purpose in view. It is a useful addition to gum-water, 1 dram to each ounce of gum present being sufficient; some copying-inks contain it. Modelling clay may be kept moist by means of glycerin.

CHAPTER IX

WATER-GLASS, LIME- AND BARYTA-WATER

THE name water-glass appears to have been first applied to those silicates of potash and of soda which are soluble in water by Professor J. N. von Fuchs, in 1825; but Glauber, so early as 1648, made a soluble potash silicate, which he termed fluid silica. Von Helmont had prepared a similar compound in 1640. The actual manufacture on a commercial scale of these salts dates, however, from 1825 only, and the credit of originating their production belongs to Von Fuchs. They differ from the compounds constituting ordinary and insoluble glass by containing no lime, baryta, alumina, or other earthy base. They are made in several ways. The purest sand obtainable is fused with carbonate of potash, or carbonate of soda, or a mixture in the desired proportions of these two carbonates, in the presence of a little powdered charcoal. The fused mass dissolves by long continued boiling in water, and yields a heavy syrupy liquid of strongly alkaline reaction. By evaporating this liquid to dryness, and fusing the residue, the water-glass may be obtained in a solid form, and then closely resembles ordinary glass in appearance. Water-glass may also be made by heating flints red-hot, quenching them in water, and then digesting the powdered silica thus obtained with soda-lye or potash-lye under pressure.

Three kinds of water-glass have been used in water-glass painting or stereochromy. One of these is a potash silicate, another is a soda silicate, the third is a mixture of these two, or a potash-soda silicate, called double water-glass. The solutions of the two former silicates as met with in commerce vary a good deal in their relative proportions of silica and alkali ; it is not desirable that they should contain so much silica as was recommended in the original papers of Von Fuchs, the inventor of stereochromy, and of Kuhlmann, who subsequently modified the process. Indeed, it has often been found useful to add a little pure caustic potash or caustic soda-solution to the commercial solutions of water-glass before diluting them with distilled water for use in this process of painting.

A solution of water-glass, if allowed to dry upon a piece of ordinary glass, leaves an opaque white irremovable stain. Water-glass alters or destroys, in virtue of its strong alkalinity, the great majority of organic pigments. On the same account it cannot be used with flake-white, aureolin, the chromates, vermilion, and several other mineral pigments. It hardens zinc-white, some of the ochres, earths, and terre verte, forming with them, or with some of their constituents, double silicates, which are quite insoluble in water. The fixative power of water-glass in stereochromy depends indeed mainly upon actions of this order which occur between it and ingredients of the plaster or painting-ground, and of the pigments. It was formerly supposed that when an alkaline silicate acted upon carbonate of lime a double decomposition occurred, of which the only products were an alkaline carbonate, and lime silicate. But subsequent investigation has proved that the change in question is more complex, a considerable quantity of a double and insoluble silicate of lime and alkali being produced. Similar double silicates of potash or soda and zinc, of potash or soda and baryta, and of potash or soda and alumina, have been proved to exist in stereochromic work; doubtless, many others are also present. They are not only insoluble in water, but are harder than the materials out of which they are formed.

Commercial solutions of water-glass contain from 28 to 60 per cent. of the alkaline silicate or silicates. They should be carefully preserved from access of air, the carbonic acid of which produces much alkaline carbonate (often separating in crystals in the case of soda), and finally causes the separation of gelatinous silica hydrates. The entrance of calcareous matters, gypsum, zinc-white, etc., should also be guarded against.

The subject of water-glass is here treated very briefly, partly because the various processes of stereochromy, even with their latest improvements, are very little used in this country, and partly because the preparations of water-glass specially made for the use of painters may be trusted. To this latter observation I might add the remark that the problem of thoroughly examining a commercial water-glass solution, for strength, purity, and due proportion of silica to alkali, is too complex to be undertaken except by a trained chemist.

Lime-water is the name given to the solution in water of slaked lime, called in chemical language hydrate of lime, calcium hydrate, and calcium hydroxide. To prepare it, quicklime, which has been made by burning (as it is commonly called) a pure marble, is slaked with distilled water. The calcium hydrate formed is placed in a wide-mouth stoppered bottle, and covered with several times its bulk of distilled water. The object of this treatment is to dissolve soda and some other soluble impurities, the major part of which will be removed when the watery liquid in the bottle is decanted from the undissolved excess of calcium hydrate, which should then be again covered with distilled water which has been recently boiled. The stopper should be well ground and smeared with vaseline. The bottle should be shaken at intervals in order that the water may take up as much calcium hydrate as it can dissolve. After all, this amount is very small, not exceeding, at 15° C., 0·172 part by weight per hundred measures of lime-water. Thus a gallon of lime-water, saturated at 60° F., could not contain more than 120 grains of calcium hydrate, corresponding to 90 grains of pure lime or calcium oxide, CaO. In ordinary practice such a perfectly saturated solution is not attainable, while the most carefully prepared and strongest solution is sure to become weakened each time the stopper of the containing vessel is withdrawn by the removal of some of the lime in solution in the form of carbonate of lime. The clearest lime-water, from this cause and from its action on glass, always appears turbid after a time.

Although so dilute a solution, lime-water gives the most marked reactions of an alkali: it turns red litmus paper blue, embrowns yellow turmeric paper, and imparts a crimson hue to colourless phenolphthalein paper. It acts energetically upon many organic and some inorganic pigments owing to its alkaline or basic properties. The ease with which the lime in lime-water unites with carbonic acid forming carbonate of lime (=calcium carbonate), and the bearing of this action, and of other properties of caustic lime upon the materials and processes of painting are discussed in Chapters II. and XXIII.

Baryta-water may sometimes advantageously replace lime-water in fresco-painting. It is a solution of hydrate of baryta, barium hydrate, barium hydroxide, for these names all belong to the compound, in distilled water. The distilled water used should have been recently boiled and then cooled out of contact with the carbonic acid of the air. The barium hydrate used may be purchased in the form of colourless crystals having the formula $BaO_2H_2 + 8$ aq. These, if not sufficiently pure, may be washed with cold distilled water, or recrystallized from boiling water, in which they dissolve very abundantly. A saturated cold solution is made by placing rather more that 1 ounce of these crystals in a bottle containing a pint of distilled water: the bottle should be almost full, the stopper should be smeared with a little vaseline. If the crystals dissolve completely, after repeated agitation, a few more should be added so as to leave a small excess at the bottom of the bottle. If the solution be clear it may be used directly from the bottle, as required; if filtration be needed, a glass plate should be placed on the funnel during the operation to prevent free access of air, and the clear filtrate should be received at once in the bottle in which it is to be preserved. A solution of barium hydrate saturated at 15° C., contains nearly 2·9 parts by weight of BaO_2H_2, in 100 measures, or 2,023 grains per gallon. It is thus very many times stronger than a solution of calcium hydrate saturated at the same temperature. Baryta-water, as it is called, is a powerfully alkaline liquid, becoming covered with a film of white barium carbonate on exposure to the air. By blowing air from the lungs through a glass tube into baryta-water, a dense white precipitate is formed.

CHAPTER X

SOLVENTS AND DILUENTS

THE liquids to which attention is directed in the present chapter are, with very few exceptions, those which are not miscible with water. Of water itself it is not necessary to say anything beyond this, that distilled water is best adapted for almost every purpose to which this liquid is applied in the preparation of pigments, and as a solvent for gum, honey, etc. Next to distilled water may be ranked rain-water collected in the open country; then the softer kinds of water yielded by streams, springs and wells. Waters containing more than 20 or 30 grains per gallon of solid matters in solution should be avoided as far as possible.

Before considering the individual solvents to which attention is directed in the present chapter, a list of the most important of those which have been obtained in a pure state may be given. These liquids are arranged in the following table according to their boiling-points, those which boil at low temperatures being placed first: an asterisk indicates that the liquid is miscible with water:

TABLE OF SOLVENTS

Name	*Boiling-point*	*Specific Gravity*	*Formula*
Ether - - -	35° C. = 95° F. -	·702 -	$C_4H_{10}O$.
Carbon bisulphide -	46° = 115° -	1·272 -	CS_2.
*Wood-spirit - -	55° = 131° -	·814 -	CH_4O.

TABLE OF SOLVENTS (*continued*)

Name	*Boiling-point*		*Specific Gravity*	*Formula*
*Acetone	56° C.	= 133° F.	·802	C_3H_6O.
Chloroform	61°	= 142°	1·525	$CHCl_3$.
*Alcohol	78°	= 172°	·794	C_2H_6O.
Benzene	81°	= 176°	·89	C_6H_6.
Toluene	111°	= 232°	·873	C_7H_8.
Pinene	160°	= 320°	·864	$C_{10}H_{16}$.
Cymene	175°	= 347°	·858	$C_{10}H_{14}$.
Dipentene	181°	= 358°	·846	$C_{10}H_{16}$.

The pinene and dipentene named in the above table are examples of what are now called terpenes. Mixtures of these and other terpenes constitute what is generally known as oil or spirit of turpentine. Terpenes are very general and very abundant constituents of the volatile, ethereal or essential oils extracted from plants. Besides the compounds included in our list and certain volatile oils, we shall have to consider some of the liquid and more volatile constituents of natural petroleum and of artificial paraffin oils. The fixed or fatty oils, which are constantly used in painting as solvents, have been already treated of in Chapter V.

Ether, often called sulphuric ether, is a very mobile liquid, having a penetrating odour, and of extreme volatility. Its vapour is given off freely at ordinary temperatures, and forms with air a highly inflammable and explosive mixture. Great care is therefore required in using this liquid; no light must on any account be brought near it. It does not mix with water, but floats on the surface, although it dissolves in water to the extent of about 10 per cent. Commercial ether always contains water and alcohol. The latter, which it is seldom necessary to remove (for varnish-making, etc.), can be got rid of only by repeatedly shaking the crude ether with water, whereby much ether also is dissolved away. The water present interferes seriously with the use of ether as a solvent for resins, etc., but it may be removed by careful rectification with fused calcium chloride, that substance having previously been allowed to remain in contact with the liquid for a day. A final distillation from a little metallic sodium completes the drying of the ether and also removes, if used in sufficient quantity, the alcohol present. Great care is necessary in distilling ether to secure, by a current of ice-cold water in the condenser, the condensation of the vapour.

Carbon Bisulphide.—This heavy, oily but volatile liquid readily gives off vapour at ordinary temperatures. It is poisonous, and the same care in manipulating it must be taken as that insisted upon in the case of ether. The smell of the ordinary commercial bisulphide is most offensive, but it is now possible to purchase a specially purified sort from which a particularly disagreeable sulphur compound of nauseous odour has been removed. Carbon bisulphide sinks in water: it is a powerful solvent for many resins, and mixes perfectly with the fixed and essential oils in all proportions.

Wood-spirit, or methyl-alcohol, is a constituent of wood-naphtha, a product of the destructive distillation of wood. It rarely occurs in commerce in a state even approaching to that of purity. It is miscible with water in all proportions, but not with fixed oils. When free from water it may be used as a solvent for some resins, and for removing discoloured varnish from oil-paintings. Methylated spirit is spirit of wine, to which some crude wood-spirit has been added.

Acetone also occurs in crude wood-naphtha. It has a penetrating but agreeable odour. It is miscible with water, alcohol, oils, etc., and dissolves many resins, camphor, fixed oils, and allied bodies. It is sometimes serviceable as a solvent for discoloured varnishes on pictures. Commercial acetone is very impure, containing wood-spirit, empyreumatic oils, and water.

Chloroform is another powerful solvent of resins. It has a pungent but sweet taste, is not miscible with water, and is very heavy. Commercial chloroform often contains alcohol and other foreign matters, from most of which it may be purified by redistillation from a little oil of vitriol followed by a second distillation from fragments of quicklime. For making varnishes neither water nor alcohol should be present in chloroform, but there are other impurities which do not interfere with its employment for such a purpose.

Alcohol, or pure spirit of wine, is met with in commerce practically free from all impurities save water. Proof spirit, rectified spirits of wine, and methylated spirit, though of service in cleaning oil-pictures and for many other purposes, ought not to be used in the preparation of varnishes. For this purpose pure alcohol, often called absolute alcohol, is required; but provided that it contains no water the presence of wood-spirit is no drawback to its use. In commerce, nearly absolute alcohol, made both from spirits of wine and from methylated spirit, is obtainable; but it may be prepared by operating upon the strongest available spirits of wine in the following manner: The spirit is distilled in a water-bath until no further strengthening of the alcoholic distillate is secured by repetition of the process; then a dry retort is half-filled with small, clean, hard fragments of quicklime, the strong spirit is poured upon these so as to somewhat more than cover them, and then the whole is left over-night; distillation from a water-bath is then commenced, when it will be found that a spirit comes over which contains no more than one part of water in two hundred. Even

6—2

this small proportion may be removed by redistilling the alcohol from a very little metallic sodium. The last distillate, when a small portion of it is shaken up with its own bulk of benzene, should mix perfectly with the latter, causing no turbidity. But it should be borne in mind that absolute alcohol is a very hygroscopic liquid, greedily absorbing water from the air; it must, therefore, be kept in well-stoppered bottles, filled almost completely. In absolute alcohol some of the more intractable resins, even some kinds of copal, readily dissolve. The specific gravity of absolute alcohol at 15° C. is ·794, while, if it contains but 1 per cent. of water, its specific gravity is distinctly higher, namely, ·797.

Benzene is employed not only as a solvent, but as a diluent of the medium or oil employed in painting. It is obtained from the lighter naphtha separated in the fractional distillation of coal-tar. The benzene (also called benzol) of commerce is rarely pure. The presence of small quantities of higher hydrocarbons of the same series is of little moment, but it also contains about one half per cent. of a sulphur compound called thiophene (C_4H_4S), to which the offensive odour of ordinary benzene is partly due. Thiophene is, however, much more soluble in cold oil of vitriol than is benzene, and may be removed by several treatments of the benzene with small quantities of this powerful acid. Benzene thus purified can now be purchased. Benzene is a mobile liquid, not miscible with water, but dissolving readily in all proportions in most if not all of the liquids now being described. It dissolves oils and very many of the harder as well as all the softer resins.

Toluene much resembles benzene, and may be used for the same purposes, although it is less volatile. Commercial toluene has a disagreeable smell, arising from the presence of a sulphur compound (thiotolene), which is more difficult to remove from the liquid than the thiophene from benzene.

Cymene belongs to the same series as benzene and toluene. It occurs naturally in Cumin oil (from *Cuminum Cyminum*), and may be prepared without much difficulty from the terpenes. It has an agreeable lemon-like odour, and may be used in painting as a substitute for oil of turpentine. It has but little tendency to resinify; it does not, however, evaporate quite quickly enough for some of the purposes of the artist.

Pinene, *Sylvestrene*, and *Dipentene*, with several other similar compounds, are the main constituents of the various liquids to which the ordinary name of turpentine, or, rather, spirit or oil of turpentine, is applied. All these liquids are hydrocarbons, having the same composition in 100 parts, expressed by the empirical formula $C_{10}H_{16}$. But these liquids—of which about ten are known—differ from one another in some of their chemical and physical characters, such as oxidizability, boiling-point, specific gravity, and action on light. The extreme importance of turpentine in the process of oil-painting, and in the manufacture of varnishes, warrants a full consideration of its several constituents.

Turpentine, properly so called, is not a liquid, but the solid or semi-solid resinous secretion of many trees, chiefly coniferous. Some exudes naturally, but much more is obtained by artificial incisions. It consists of a mixture of one or more true resins in which oxygen is present, with one or more liquid hydrocarbons which contain (as the name imports) nothing but carbon and hydrogen, and therefore no oxygen. These hydrocarbons are called in chemical language *terpenes*, a term by which they will be designated henceforth in the present chapter. On distilling the crude turpentine or resins alone or with water, or in a current of steam, the terpenes distil over while the solid part remains behind; this, on fusion, is called rosin or colophony. It need not be further considered, as it is of no value in painting, being friable and more or less strongly coloured. We confine our attention, therefore, to the distillate or terpenes. It should be added, however, that the leaves, cones, and other parts of many coniferous trees, themselves yield various terpenes when submitted to distillation, and that many of the volatile or essential oils of aromatic plants other than conifers contain or consist of terpenes. The oils expressed from the rinds of lemons and oranges, or obtained by the distillation with water of the flowers of lavender, afford illustrations of this remark.

Terpenes differ from one another in several obvious and in several obscure ways. Even now the chemistry of these liquids is not by any means clearly and completely unravelled. We need not here concern ourselves with those minute differences in chemical and physical properties by which the identity of individual terpenes is established, but may confine our attention to their most salient characteristics. Of these none is more important than the behaviour of terpenes with regard to atmospheric oxygen. Some of these liquids absorb oxygen readily, and to a large extent, from the air, becoming thereby resinified—in fact, they thus yield sticky, resinous, semi-solid bodies, closely resembling the crude turpentine from which they have been prepared. Everyone who has had occasion to use spirits of turpentine frequently must have noticed the production of a sticky substance about the neck of the bottles in which this liquid has been kept. Moreover, the spirit of turpentine itself will often have been noticed to have become cloudy, viscid, or almost solid, especially if it has been contained in a bottle frequently opened, and not quite full. Besides these observations another will have been made—different specimens of spirits of turpentine will have been found to differ much as to the rate at which these changes have taken place. Some samples, even in half-full bottles, remain clear and limpid for long; others become thick, opaque, and sticky in a few weeks. Such changes are undesirable in a solvent, diluent, or painting medium, on many grounds. The resin formed is an unsatisfactory one—soft, sticky, and contractile. The liquid decreases so greatly in mobility, and increases so greatly in viscosity, that its utility in thinning oil pigments, and in making fine touches, is greatly impaired. And this thickening of the liquid is accompanied by the production of acid substances and of water, which affect injuriously the ease of working and the stability of the picture. Spirits of turpentine should disappear by evaporation quickly and completely from the painting into which it has been introduced. Now, if it be easily oxidized, even if it be kept from experiencing change *before* it is actually employed, it will, during the very time in which it is being used, attract oxygen; so that though a great part of it will escape by evaporation, the remainder will resinify on the very canvas itself, adding a sticky deposit to the drying oils and hard resins which may have been used as the painting medium. It is clear, from all the above considerations, that the greatest care ought to be taken in selecting, in the first instance, such a sort of spirits of turpentine as will resist oxidation under ordinary conditions. Even an inferior spirit may be used, with a minimum of disadvantage, if immediately after distillation it be poured into a number of small bottles, so as to fill each of them completely; they should be at once closed with sound corks. In this way the contents of a bottle may be used up very soon after it has been opened. Another precaution may be taken: A few small fragments of hard quicklime may be placed in each bottle to absorb any moisture produced by oxidation, and also the acid bodies which are formed at the same time. Even with the choicer samples of spirits of turpentine, which pass much less easily into resins, this use of quicklime is desirable; but in this case the employment of many small bottles is unnecessary, and it will suffice to put a few hard pieces of lime, free from powder, into a pint or quart bottle, and then to fill it with the spirit. The clear liquid may be poured off as required for use, any disintegrated particles of lime sinking readily to the bottom of the vessel.

Before giving details as to the sources and characteristics of the best terpenes, it may be useful to mention that commercial samples of spirits of turpentine may be tested and compared by means of a very simple experiment. Obtain the required number of small, flat-bottomed, conical glass flasks with wide mouths, one flask for each sample; these flasks are known as Erlenmeyer's. Into the flasks pour enough of the several samples to cover the bottom to the depth of one-eighth of an inch; label each flask to correspond with the sample, and lightly close each mouth with a plug of carded cotton — the date of the experiment should be added on the label. Shake each flask so as to cause a number of bubbles to be formed in the liquid; the more rapidly these bubbles break, the better is the sample. Repeat the experiment of shaking the samples at short intervals for a few weeks—notable changes in the viscosity of the oils will be observed sooner or later. Any sample which after one month remains clear, and in which the bubbles formed on agitation break almost as quickly as at first, may be accepted as of good quality. Another test for discriminating between the samples, so far as their state at the time of the experiment is concerned, is the very simple one of placing one drop of each oil upon a sheet of writing-paper, and gently warming the translucent stain it forms; with a good oil the mark completely disappears.

Two other obvious characteristics of different samples of spirits of turpentine may now be noticed—namely, odour and boiling-point. Some samples have a much more agreeable scent than others; the vapour of these seems to have a less marked tendency to produce headache than that of the pungent and cruder-smelling varieties. The range in boiling-point is not very extensive; but it may be taken as about 25° C., the figures ranging from 155° to 181°. Samples having lower boiling-points evaporate more quickly than those which enter into ebullition at higher temperatures; each type will have its appropriate use in the process of painting. The solvent power on resins differs with different kinds; this is a property which is of importance in varnish-making, but very little accurate knowledge exists on this point.

A few of the more important turpentine oils and essential oils containing terpenes may now be named:

Russian and Swedish oils, chiefly from *Pinus sylvestris*.

Austrian oil, partly from *Pinus Laricio*, partly from *P. Pumilio*.

American oil, chiefly from *Pinus australis* and *P. tæda*.

French oil, from *Pinus Pinaster*.

German oil, from *Pinus sylvestris*, *P. Cembra*, *P. Abies*, *P. vulgaris*, etc.

Strasburg oil, from *Abies pectinata*.

Juniper oil, from *Juniperus communis*.

Eucalyptus oil, from *Eucalyptus globulus* and other species.

Orange oil, from *Citrus Aurantium* and its varieties.
Lemon oil, from *Citrus medica*, var. *Limonum*.
Spike-lavender oil, from *Lavandula spica*.

The above-named turpentine oils and fragrant essences are accompanied by various resins, camphor, and other oxygenated bodies, from which they may be separated by treatment with caustic potash, metallic sodium, and fractional distillation. The oil of spike, however, yields a comparatively small amount of terpene, although more than English oil of lavender, which is distilled from the flowers of another species of *Lavandula*, *L. vera.* How far the liquid oxygenated compounds present in eucalyptus oil and spike oil are in themselves valuable as diluents and solvents has not been ascertained, and these oils are, in fact, used for painting purposes without any preliminary treatment save that of redistillation.

From the above-named liquids a number of terpenes have been isolated. Among the better known of these the following may be mentioned.

1. *Pinene*, with a boiling-point of 160° C. It forms the chief constituent of German and American oil of turpentine, and occurs in considerable quantity in the volatile oils of wormwood, sage, rosemary, and juniper. It is an easily alterable and resinifiable terpene, and should not be employed either in thinning paints, or as an ingredient of varnishes.

2. *Phellandrene.* — Boiling-point 171° to 172°. This terpene has been separated from eucalyptus oil, that is, from the oil obtained by the distillation with water of the leaves of one of the numerous species of eucalyptus, *E. amygdalina*: whether it occurs in all eucalyptus oils has not been ascertained. It is one of the most alterable of all terpenes, and the oils containing it should be avoided.

3. *Limonene.*—This terpene, like most of the others, occurs in two forms or varieties, having opposite actions on polarized light. It is sometimes called citrene. It boils at 175°. It is best prepared from orange-peel oil, which yields over 80 per cent. of dextro-limonene when distilled from caustic potash. When pure it is less alterable than pinene and phellandrene, though it resinifies after a time.

4. *Sylvestrene.*—Boiling-point 176° to 177°. When pure it has the smell of bergamot, but generally presents the odour of fir-wood. It is dextrorotatory, and forms the chief constituent of Russian and Swedish oil of turpentine, and of some of the German oils. On exposure to air it leaves a sticky resin.

5. *Dipentene*, which boils at about 181° C., is optically inactive, and may be made by heating some of the other terpenes to 250°—270° for some hours, or by mixing dextro- and lævo-limonene together. The only natural product in which dipentene is known to occur is the volatile oil which accompanies common camphor. Its odour resembles that of citron oil: it resinifies to about the same extent as limonene.

From the preceding descriptions it may be gathered that of all the above terpenes limonene and dipentene are the least alterable. They have, however, somewhat high boiling-points, and evaporate more slowly than most of the other terpenes.

A few more words may be added in reference to oil of spike-lavender. The terpenes present in this agreeably-smelling liquid sometimes amount to 4·0 per cent., part at least of the remainder consisting of liquids in which oxygen as well as carbon and hydrogen occur. It is a powerful solvent for many resins, and forms a part of Gambier-

Parry's spirit-fresco medium. It should not be redistilled from caustic potash, but from a few lumps of quicklime, which have been in contact with it for twenty-four hours. And here it may be mentioned that the presence of water in a terpene, or a mixed essential oil, may be detected by the cloudiness which it shows when mixed with thrice its volume of benzene or of petroleum spirit. To remove traces of water from any of the less volatile liquids we have been considering, without having recourse to distillation from caustic potash, or from quicklime, the following simple procedure may be adopted: A glass flask is three fourths filled with the liquid, and then it is kept at a temperature of 110° to 120° C., so that the moisture present is disengaged as vapour without the terpene or essential oil itself boiling: drops of moisture will condense in the neck of the flask, and may be removed from time to time by means of a roll of blotting-paper. The mouth of the flask should be loosely plugged with carded cotton. Of course this process is applicable only to liquids which boil at temperatures considerably over 120°, like the terpenes.

In connection with the terpenes two other liquids and one solid remain to be mentioned. The liquids are 'oil of amber' and 'oil of copal.' These are obtained by strongly heating the resins in question. They are employed as efficient solvents for the harder resins. Oil of amber may be obtained in commerce at a moderate price. Its offensive smell may be partly removed by adding to it some white lead and solid caustic potash, and afterwards distilling it. It contains amongst other liquid constituents at least one terpene. Its boiling-point rises, as distillation proceeds, from 110° C. to 260°. Camphor is expressed by the formula $C_{10}H_{16}O$, and is obtained chiefly from *Cinnamomum camphora*, a tree of China and Japan. It is a tough crystalline solid of penetrating odour and pungent taste. It is soluble in all the liquids named in the present chapter. Although it boils at so high a temperature as 204° C., it readily and rapidly volatilizes at ordinary temperatures. It is used to aid the solution of some of the harder resins in the making of varnishes, but its presence in a varnish is objectionable, for it slowly escapes after the apparent drying-up of the varnish, and thus causes a deterioration of the lustre and continuity of the resinous film.

Petroleum-spirit.—When native petroleum and the similar materials obtained in the distillation of bituminous shales, etc., are submitted to fractional distillation, the more volatile portions which come over first constitute the liquids variously known as gazoline, benzoline, ligroine, petroleum-naphtha, petroleum-ether, and petroleum-spirit. This liquid consists entirely of hydrocarbons, some of which belong to the paraffin series, while others are naphthenes. Their boiling-points are all under 170° C., while some of them boil as low as 50°; indeed, commercial samples of petroleum-spirit often begin to enter into ebullition at a lower temperature even than this. The series of petroleum products may be roughly grouped thus:

Petroleum-spirit boils below 170° C.; specific gravity, 0·6 to 0·7.

Lamp-oil, kerosene, photogen, or paraffin-oil, boils between 180° and 220°, and has a specific gravity of 0·78 to 0·82.

Solar-oil, lubricating-oil, vaseline, and paraffin-wax, are heavier products, with a range of specific gravity from 0·83 up to 0·94. Their viscosity increases with their density until the semi-solid vaseline and the solid paraffin-waxes are reached. The latter substances have been described already, the former are not available in painting: in fact, their presence even in traces in petroleum-spirit—an extremely useful solvent and diluent—should be carefully guarded against. They neither escape by evaporation nor harden in the lapse of time Thus petroleum-spirit remains alone for further consideration.

As a solvent for resins, and as an extremely volatile and very *thin* liquid for diluting oily vehicles and paints in the process of oil-painting, the variety of petroleum-spirit which boils between 50° and 70° C. is the most suitable. It contains hydrocarbons represented by the formulæ C_6H_{12} and C_6H_{14}. It must be used with great caution on account of its easy inflammability and the readiness with which it gives off a vapour, which, when mingled with atmospheric air, is highly explosive. It may be used for many purposes in lieu of benzene (from coal-tar naphtha), being much cheaper and quite as efficient. A drop of this variety of petroleum-spirit on paper evaporates very quickly, leaving no greasy stain.

Another variety of this petroleum-spirit is obtained by collecting apart the fractions which boil between 100° C. and 130°. These contain heptane (C_7H_{16}), octane (C_8H_{18}), heptylene (C_7H_{14}), and octylene (C_8H_{16}), and other hydrocarbons. This mixture is less volatile than that just described, it dries more slowly, and is a less energetic solvent.

A third variety boils between 130° and 170°, and is available for many of the purposes for which turpentine-oil is employed. It is not advisable, in my opinion, to use fractions having a higher boiling-point than 170° C. as additions to the pigments and vehicles of oil-painting, for, though their slow drying is sometimes an advantage, there exists the danger of their incomplete evaporation from the painted surface. If they remain even in traces in the finished work after it has been varnished, they may give rise to the same accidents as are caused by the treacherous though seductive asphaltum.

It should be remembered that the various petroleum liquids just described do not resinify, nor do they leave any permanent stain or mark upon paper which has been moistened with them.

CHAPTER XI

SICCATIVES OR DRYERS

THE terms 'siccatives' and 'dryers' are applied to three classes of substances. Perhaps the most correct or appropriate application of these words is to those metallic compounds which are used in order to increase the rate at which the drying oils harden, but in the literature of the subject we often find that drying oils which have been thus treated, and likewise certain resinous solutions, are spoken of as siccatives. In the present chapter we describe the dryers proper only, referring our readers to the chapters on oils and on varnishes for the necessary particulars concerning the other materials which may be included in the group under discussion.

Lead and several of its salts have been long and widely used as dryers. Metallic lead in the form of foil, litharge or lead protoxide, minium or red lead, lead peroxide, sugar of lead or lead acetate, the basic lead acetate, and white lead itself, have all been used in this way, chiefly for the purpose of making linseed or other painting oil dry more quickly. Some of these compounds, particularly sugar of lead, have been introduced into the very picture itself. It was a common practice to employ powdered sugar of lead or a solution of this salt in water to hasten the drying of vehicles and of slow-drying pigments which had been ground in oil. I have seen one of the results of this commingling of sugar of lead with the medium or the paint in the production of an immense number of small spots in the picture, sometimes appearing through the surface-varnish in the form of a white efflorescence. This efflorescence consists at first of lead acetate in crystals, but these soon attract carbonic acid from the air and become lead carbonate, which, in its turn, is changed into lead sulphide by the action of sulphuretted hydrogen. This tendency of the lead compounds to yield brown or black lead sulphide is, indeed, the great drawback to any use of these substances as dryers. When oil is left in contact with them, and especially when heat is applied to the mixture, some of the lead dissolves, forming, with the fatty acids of the oil, lead-soaps. These soaps are distributed uniformly throughout the oil, and help to make it dry and harden quickly. The same action occurs when white lead is ground as a paint with oil, and has been urged as an objection to the use of those white leads which contain hydrate of lead, a compound which acts upon oil more quickly and thoroughly than the carbonate of lead. It will be seen, however, that while there may be reasons for permitting the use of a single lead pigment which possesses this peculiar property, there can be none for introducing into every part of a picture oils or other materials which contain a metal, like lead, so liable to cause discolouration and darkening, when other and perfectly innocuous substances are available for producing the same siccative effects. On this account we omit further reference to the lead compounds, which have been and are still employed in the preparation of strongly-drying oils, etc., but pass on to the

Manganese compounds, of which the binoxide, the

hydrated protoxide and sesquioxide, the borate, the oxalate, and the linoleate are the most important.

Manganese binoxide, the black oxide, MnO_2, is used in the form of a powder obtained by grinding the mineral pyrolusite. As the effectiveness of this compound is made complete only by the use of oil of vitriol, which needs subsequent neutralization with lime, it cannot be recommended as a material for rendering linseed oil intended for painting, or for making picture-varnish, more drying. The difficulty of preparing the manganese hydrates above mentioned constitutes an objection to their employment for this purpose. But the borate, the linoleate, and the oxalate of manganese may be obtained in commerce in a state of sufficient purity for our present purpose, and it is to them that we wish to direct attention. Borate of manganese may, moreover, be so easily prepared, that it is worth while to give here the necessary directions. One pound of pure manganese sulphate is dissolved in six pints of distilled water, the solution being filtered if cloudy. A few drops of the liquid are now to be tested with caustic soda solution—the precipitate formed should be white; if it show a greenish, yellowish, or greyish hue, iron is probably present, and it will be necessary to treat the whole of the solution with caustic soda until a white precipitate falls, and then to filter it again. In order to produce manganese borate, a boiling saturated solution of pure borax is added to the manganese sulphate solution until no more precipitate falls. The precipitate is collected on a filter and washed with hot distilled water until the wash-waters show no turbidity when a solution of barium chloride and a few drops of dilute hydrochloric acid are added to the last portion coming through the paper. The borate of manganese is then dried in a warm place, and finally in the water-oven.

One grain of it, warmed with linseed oil, is sufficient to render an ounce of the latter highly drying (see Chapter V.). The oxalate or the linoleate of manganese may be used in the same way.

Borate of lime and borate of zinc have been employed not only for rendering oils more quickly drying but also in admixture with some of those oil-paints which dry with difficulty. Being colourless they are well adapted for use with white pigments, such as oxide of zinc. Several of the siccative materials sold under various fancy names consist of mixtures of these borates with carbonate of zinc or oxide of zinc, manganese compounds being also sometimes added. Another dryer in common use is white vitriol or sulphate of zinc. This is used either in the form of the hydrated salt, which contains 7 molecules of water ($ZnSO_4$, $7H_2O$), or less after *gentle* heating, by which some of the water is driven off. It is, however, less effective and advantageous than the manganese compounds above recommended. Most of the other siccatives employed by artists owe their efficacy to lead, or are resinous preparations. Siccatif de Courtrai is a very dangerous mixture, heavily loaded with compounds of lead; Siccatif de Haarlem is a resinous preparation, which produces, on drying, a hard, brilliant, and tough film.

CHAPTER XII

VARNISHES AND VEHICLES

When an oil, such as linseed, walnut, or poppy, has been purified and made more quickly drying by one or other of the methods already described, it is often called 'varnish.' It has acquired the property of rapidly solidifying, when spread as a thin layer, into a tough transparent substance, endowed with a considerable degree of cohesiveness and elasticity, yet rather soft withal. Now oil of this character, although it has many uses in painting, is not quite hard enough for some of the purposes for which a true varnish may be required, but its defects may be amended by associating with it one or more of the resins described in Chapter VI. One class of varnishes is compounded in this manner of two materials, oil and resin, both of which are fixed or non-volatile. A second group of varnishes consists of a resin dissolved in a volatile solvent. And there are also mixed varnishes which contain at least three ingredients—namely, a drying oil, a volatile solvent, and a resin. As the varnishes which consist wholly of oil and resin are thick and intractable, it is usual to thin these, according to the purpose for which they are intended, with varying amounts of some volatile liquid or solvent, spirits of turpentine being most frequently thus employed. In order to avoid too elaborate a classification, it will be advisable to describe those varnishes which contain oil as *oil or fat varnishes*, and those which consist wholly of a resin and a volatile solvent as *spirit-varnishes*. We describe the latter first, as their manufacture is easier and their constitution simpler. In order to avoid repeated references to the descriptions already given of the several materials employed in making varnishes, it will be convenient to state once for all that the oils used are described in Chapter V., the resins in Chapter VI., and the solvents in Chapter X.

Mastic Varnish.—This is usually prepared by dissolving mastic in spirits of turpentine, although other volatile oils and even absolute alcohol may be employed. In order to prevent the mastic from agglutinating together, warm powdered glass, or warm fine white quartz sand, may be added to the mixture of resin and solvent. The spirits of turpentine should be absolutely free from moisture, the mastic may be in tears, or, preferably, have been purified and dried as before directed. The materials are introduced into a capacious glass flask fitted with a cork, tube and condenser so arranged that, when the flask is heated in a water-bath, the vapours given off from the solvent may be condensed and return to the vessel. The temperature of the water-bath may be 100° C. if oil of turpentine be used, but should not be allowed to rise beyond 80° C. if absolute alcohol or 96 per cent. alcohol (specific gravity ·806) be substituted for the oil of turpentine. The following receipt gives a varnish which contains nearly 25 per cent. of its weight of mastic, but the proportion may easily be increased or diminished :

14 ounces of mastic,
44 ,, ,, spirits of turpentine,
6 ,, ,, powdered glass, or fine sand.

When the mastic has dissolved the varnish is allowed to cool, and then poured off into a closed glass vessel, in which it is allowed to rest until perfectly clear. Or it may be clarified by filtration through a plug of *dried* carded cotton fitted into a funnel. The funnel should be closely covered with a ground glass plate, but a specially contrived filtering apparatus has been designed for the purpose of preventing any escape of vapour during the process of filtration.

The varnish prepared according to this receipt is nearly colourless, and leaves a brilliant glassy film when it evaporates on a smooth surface. But this film is very brittle, and easily abraded by gentle friction even with the finger, in fact it consists of little more than the original mastic resin, the fragility of which is well known. To obviate this brittleness many plans have been devised. Sometimes Venice turpentine, Canada balsam, or Elemi resin is introduced in small quantity, not exceeding one-seventh in weight of the mastic used. In consequence of such admixture of a natural soft turpentine the varnish produced dries more slowly, and leaves a less brittle, tougher, more adhesive, and more elastic film on evaporation. Ultimately, however, these balsams become brittle like mastic itself. This remedy is, therefore, of a temporary character, but, at the same time, these additions do not interfere with the ease with which the varnish, when old and discoloured, can be removed from a painting by means of solvents or of friction, without injuring the glazing pigments which may lie immediately below it: they also render the varnish more easy of application. The other classes of substances added to toughen the resinous film left by the drying of a spirit varnish, are fixed oils, and those liquid paraffins which boil at temperatures above 170° C. A very small proportion of 'manganese' linseed oil is, perhaps, the more effective and safer toughener of the two, but its introduction involves the disadvantage just named. In many French mastic varnishes camphor is introduced for the same purpose to the extent of 5 to 8 parts for each 100 of mastic. The camphor, however, gradually escapes by volatilization, the varnish losing its fine lustre and becoming brittle and fissured. It should be mentioned here that the more easily resinified varieties of oil of turpentine, when used as solvents for mastic, also toughen the resinous film left on the drying up of the varnish, although the effect is not permanent. If alcohol, benzene, light petroleum ether, or other non-oxidisable solvents be substituted for any kind of essence of turpentine in making mastic varnish, there is no doubt that the brilliant films they yield are more brittle and less adhesive

Sandarac and the various kinds of soft pale dammar may be substituted wholly or in part for the mastic mentioned in the receipt for spirit varnish above given But if these dammars be used great care must be taken that they are themselves free from moisture, and that the oil of turpentine or other solvent be also perfectly dry. It has been recommended to employ oil of spike lavender instead of oil of turpentine in making mastic varnish. The spike oil in this case must be free from water, and freshly distilled: mastic varnish thus prepared has less tendency to 'bloom' than the ordinary kind, but if pictures are varnished in a perfectly dry atmosphere and kept therein till the surface has hardened, the formation of bloom is minimized.

A copal spirit varnish may be made by the use of acetone, or of ether (both water-free), or of absolute alcohol, light petroleum-ether, or benzene. The copal to be dissolved may be either Sierra Leone copal, Zanzibar copal, or Demerara copal, the first two yielding the harder varnish, but the last-named being easier of solution. The powdered copal, prepared as directed previously (by exposure to the air, and heating), or first fused, or at least heated till it has lost from 10 to 20 per cent. of its weight, is kept in contact with four times its bulk of the solvent until it is nearly dissolved. Three measures of dry oil of turpentine are then added, and the mixture submitted to distillation from a water-bath until three measures of the acetone or other original solvent have been drawn over: an efficient condenser must be used. If it be desired to prepare a *mixed* varnish (partly oil or fat varnish) 1 measure of 'manganese' oil, and two measures of oil of turpentine may be used in lieu of the quantity of turpentine above mentioned, the distillation being then proceeded with as before.

In another method of preparing copal (and amber) spirit varnishes the resins duly prepared and powdered are heated with the selected solvent under pressure—that is at a temperature above that at which the particular solvent used boils under ordinary conditions. With purified oil of amber, oil of copal, oil of turpentine, oil of spike, or the heavier petroleum spirit, and on a small scale, glass tubes hermetically sealed and heated to 200° C. may be used, but if a higher temperature or more volatile solvents be employed copper tubes with screw stoppers are necessary. But operations of this order can be carried out safely and successfully only in a well-equipped laboratory or factory by skilled operators, and it is therefore unnecessary to furnish further particulars in a work like the present.

The preparation of fat or oil varnishes with the harder resins is generally attended with considerable difficulty; but there is, as we have already mentioned, one way in which the difficulty may be lessened. By the aid of one of the powerful and very volatile solvents previously named, we prepare a spirit copal or amber varnish: we then add the required amount of 'manganese' oil and draw off the volatile solvent by distillation, thinning the resinous solution obtained with so much oil of turpentine as is necessary. If the copal or amber employed has been first roasted or fused, the varnish produced will be more or less dark in tint; it is on this account that the exposure of the powdered resin to the air in a flat porcelain dish for seventy-two hours, at a temperature (of 220° C.) which does not cause discolouration, is recommended. But if, on the other hand, the copal or amber be merely powdered, some part of it, and that a considerable part, will probably remain undissolved though swollen, and will therefore be wasted. The following process, in the main identical with one recommended in the American edition of Mr. Erwin Andres' work on varnishes, yields a pale and durable varnish when Sierra Leone copal or other hard copal is employed, and is doubtless well adapted for the preparation of amber varnish also. As will be seen, it is based upon the preliminary partial solution of the hard resin in chloroform, or in light petroleum spirit of about the same boiling-point. It may be stated at once that the proportions of the five ingredients used are approximately 10 parts by weight of copal or other hard resin; 5 parts by weight of dried powdered glass or sand; enough chloroform to cover the above substances; 35 parts by weight of oil of turpentine, and 10 parts by weight of 'manganese' oil. The following is an outline only of the process. The copal, after having been powdered and heated to 220° C. for seventy-two hours, is mixed with the glass or sand, and introduced into a retort; chloroform in quantity sufficient to cover the mixture is added. After the lapse of twenty-four hours the dry oil of turpentine is poured in, and an upright condenser is attached to the retort. The retort is then heated to 50° or 60° C. for two hours, so that the chloroform continually returns to the mixture. Then the contents of the retort are allowed to cool, and the condenser slanted downwards to allow of the chloroform being distilled over. This removal of the chloroform having been effected at a temperature so low that very little turpentine has come over, the remaining mixture in the retort is heated once more with the condenser in an upright position. The heat used must suffice to bring the oil of turpentine into vigorous ebullition—in an hour the whole of the copal should have dissolved. The 'manganese' oil wanted should now be heated to 100° C., and then the copal mixture, when it has cooled to 70° C., added little by little to it with constant stirring, the temperature of the oil being maintained at 90° to 100°. When the mixture is complete the source of heat is withdrawn, but the varnish is still stirred for twenty minutes. Then it is allowed to settle, until quite clear, in glass bottles, or, if an appropriate filtering apparatus is available, it is filtered. In the latter case a little hot oil of turpentine may be used to extract any copal solution which may remain with the powdered glass or the sand in the retort.

The older plan of preparing oil or fat varnishes with hard resins is still that usually adopted; but it yields products which are darker in colour than those obtained by the method just described, as the copal or amber used has been previous heated or even fused, whereby it has lost one quarter of its weight. One way of carrying out this plan consists in melting the copal in one vessel, and heating the oil until it commences to give off small bubbles in another; then half the oil is poured in a very thin stream into the melted resin, and incorporated therewith by constant stirring. Complete union having been effected between the two materials, the mixture is incorporated with the remainder of the hot linseed oil, any portions adhering to the vessel being afterwards dissolved by means of oil of turpentine; 30 parts of melted copal, 100 parts of linseed oil, and 70 parts of oil of turpentine, are proportions often employed in carrying out the process we are describing. This process may now be completed by adding to the solution of copal in linseed oil 3 parts of litharge in fine powder, or $\frac{1}{4}$ of a part of manganese borate, stirring continually, and heating for two hours, or until the solution has acquired the character of a thick gold-coloured syrup which can be drawn out into threads. This point having been reached, the heating is discontinued, and the contents of the boiler allowed to cool to 60° or 70° C., and then is added the warm oil of turpentine which has been used to dissolve out any of the copal solution clinging to the vessel in which that resin was melted. Finally, the remainder of the oil of turpentine is very gradually introduced with constant stirring. Copal varnish prepared in the above manner ought to dry in twelve hours or sooner. It is scarcely necessary to say that this method of preparing varnish with copal or other hard resin is one that no inexperienced person should attempt; not only is there some chance of partial or total failure, but there is serious risk of fire. An easier and less dangerous process requires a specially constructed heater, which is kept hot by a water-bath. Melted copal, copal or amber oil, 'manganese' oil, and oil of turpentine, are the materials used. They are all introduced together, and, as the temperature during the process of cohobation does not exceed 100° C., the time required is greater than in the previously described process.

A good copal or amber varnish ought to leave a film (on a sheet of glass) which combines the qualities of hardness and toughness. The toughness is given by the oil, the hardness by the resin. Such a film should not become fissured even when it has been exposed to sunshine during a year. Much of the copal varnish of commerce is not made from true copal or animé at all, kowdi or kauri resin (from Dammara australis), which is much easier to dissolve, being employed instead—the product, however, is decidedly inferior. Sometimes several resins are mixed together in the preparation of a so-called copal varnish. A guarantee of genuineness, in which the name of the resin employed is inserted, should always be demanded when buying copal varnish. This ought to be furnished by the varnish-maker himself, for artists' colourmen rarely prepare oil-varnishes themselves.

For the general use of painters in oil nothing more is wanted than true copal or amber oil-varnish, a drying oil, and a diluent. Of these three liquids a mixed medium in general use is compounded by taking equal measures of the three—varnish, oil, spirits of turpentine—and mixing them together in small quantities as required. But considering the large quantity of oil already associated with oil-pigments and present in copal or amber oil-varnish, one-third of oil in the medium seems a somewhat high proportion. I have proved by numberless experiments that it may be reduced with perfect safety to the permanence of the picture, although the manipulation and technique of a painter may demand the peculiar quality in a medium which oil in considerable proportion can alone supply. A formula which answers well is this :

2 measures of copal oil-varnish made from Sierra Leone or Zanzibar copal ;
1 measure of poppy oil ;
2 measures of oil of turpentine or oil of spike.

By substituting linseed oil for the poppy oil a more quickly-drying medium is obtained; still more rapid drying is secured by means of 'manganese' oil. With the same object in view, benzene may be used instead of oil of turpentine. This latter ought, of course, in all cases, to be the one of the least resinifiable varieties obtainable.

If an amber or copal varnish containing no oil be available, a good medium may be prepared by mixing 3 measures of it with 2 measures of poppy or linseed oil, and 1 measure of oil of turpentine or of spike.

Bell's medium contains no resin, but consists of thickened linseed oil dissolved in oil of spike. To prepare it, pure linseed oil may be oxidized until it has acquired the consistence of fresh honey: this change occurs when a layer of oil is exposed to the air in a large flask, the mouth of which is lightly plugged with carded cotton. The flask should be shaken occasionally to mix its contents, and to prevent the formation of a pellicle on the surface of the oil. Paintings executed with this thickened oil medium do not acquire the hardness and solidity of those carried out with a vehicle containing a hard resin.

Of megilp—a mixture of linseed oil and mastic varnish—it is only necessary to say this: that however agreeable as a medium with which to work, it contains a poor and weak resin, which becomes in course of time yellow and brittle, and is liable to be injuriously affected when a picture, in which it has been used freely, is *cleaned.*

A good medium, consisting of copal oil-varnish and poppy oil with a trace of white wax, is a favourite amongst some artists, and may be recommended as perfectly safe.

For painting in oil on plaster, slate, or stone, a perfectly sound and convenient medium is made by warming 12 ounces of oil of spike, or of non-resinifiable oil of turpentine in a glass flask plunged in water heated to the boiling-point, and then pouring into it in a slender stream 4 ounces by weight of paraffin-wax (melting-point 62° C.). The mixture becomes perfectly clear if it be thoroughly agitated and maintained at a temperature of 80° C. Then 20 measured ounces of 'picture'-copal varnish, or 16 ounces of oil-copal varnish, are slowly added, with constant shaking, in the same way. The 'paraffin-copal' medium thus obtained may be diluted with oil of turpentine exactly to the same extent as recommended by the late Mr. Gambier-Parry in the case of his 'spirit-fresco' medium, and may be used in the same way and for the same purpose. Paintings executed with this medium present a perfectly dead or matt surface without the least *shine.* This medium is superior to that used in spirit-fresco, for it contains neither elemi-resin nor wax, the two doubtful constituents of the latter preparation.

Mr. Gambier-Parry's medium, to which reference has just been made, is prepared with five ingredients. The original instructions are unnecessarily complicated, and may be simplified while keeping to the original proportions, and without modifying the nature of the product in the slightest degree. Eight ounces of oil of spike are warmed in a glass flask to 80° C., then 2 ounces by weight of elemi are added, the mixture being warmed and shaken till the elemi has dissolved. Some dirt and woody fragments are sure to be introduced with the elemi, and so the solution (still warm) must be filtered. Upon the filter, when all the liquid has run through, 2 ounces by measure of oil of turpentine, heated to 80° C., are now poured, and the united filtrates are thoroughly mixed. The liquid is then introduced into a flask, and heated to 80° C.; then 4 ounces by weight of pure white wax (previously melted) are poured in a thin stream into the solution of elemi and thoroughly shaken.

When the commixture is complete, 20 ounces by measure of 'picture'-copal varnish, or 16 ounces of oil-copal varnish, are gradually introduced with constant agitation. The water surrounding the flask is now made to boil, and kept boiling for five minutes. The flask is withdrawn, wiped dry, and allowed to cool. As the cooling proceeds the flask is gently agitated from time to time. When the mixture begins to get treacly in consistence it is at once poured into the bottles (bottles with wide mouths, holding 4 ounces apiece, are convenient) in which it is intended to preserve the medium for use. The dilution of this medium and the mode of using it are described in Chapter XXIII. on Painting Methods.

PART III

PIGMENTS

CHAPTER XIII

WHITE PIGMENTS.

FLAKE-WHITE: White Lead—Céruse—Blanc d'Argent—Blanc de Plomb—Kremser Weiss.

WHITE lead was known to the ancients. A face-powder or cosmetic, found, in its original pottery-box of about B.C. 400, in the neighbourhood of Athens, proved to be a mixture of white lead and whitening. Theophrastus, Pliny, and Vitruvius describe its manufacture from lead and vinegar. It was designated by several names, such as cerusa, cerussa, cerosa, psimuthion. In the first half of the fourteenth century it is mentioned as 'minium album.' It has been called by divers names after the place or method of its manufacture, or after the name of persons who have devised special processes for preparing it.

White lead still continues to be made for the most part by processes which are essentially identical with the old method, now generally known as the 'Dutch' process. This consists in attacking metallic lead, in the form of 'crates,' 'grids,' or spirals, simultaneously by acetic acid, carbonic acid, atmospheric oxygen, and water-vapour. The metal is gradually converted into a mixture or compound of lead carbonate and lead hydrate. Other processes, generally yielding an inferior product containing more carbonate and less hydrate, have been used. One of these consists in passing a current of carbonic acid gas through a solution of lead subacetate; in another, 4 parts of litharge, 1 part of common salt, and 16 parts of water, are kept in contact for some hours with constant agitation, and then carbonic acid gas is led into the mixture until it becomes neutral to test-papers.

The best white lead contains two molecules of lead carbonate intimately associated with one molecule of lead hydrate, and is represented by the formula $2PbCO_3, PbH_2O_2$. This formula corresponds to about 70 per cent. of lead carbonate, and 30 per cent. of lead hydrate. If the proportion of hydrate rise above this percentage, the opacity of the paint is lessened seriously; if it fall much below the above-named figure, the binding-power and working quality of the white lead are impaired. It is scarcely necessary to say that the metallic lead used in the manufacture should be as nearly pure as possible, such, for instance, as the lead from the Upper Hartz, which contains but 2 parts of foreign metals per 1,000. These foreign metals, the presence of any one of which in sensible quantity may cause a discolouration of the product, are copper, bismuth, silver, cadmium, antimony, nickel, and, more particularly, iron. But not only must the raw material be pure, but it is necessary to guard against the contamination of the white lead during its manufacture by dust or sulphuretted gases.

The impurities and defects of white lead are (1) accidental, (2) intentional.

(1) Of the accidental impurities and defects of white lead made from pure metal, the following are the chief:

a. Metallic lead, imparting a gray hue to the product.
b. Massicot or litharge, the yellow oxide of lead.
c. Minium or red lead, which gives a rosy hue.
d. Excess of lead hydrate, which causes translucency.
e. Excess of lead carbonate.
f. Lead acetate.

A simple experiment will suffice to show whether lead acetate be present in objectionable proportion in any sample. Some of the dry pigment is to be ground with distilled water into a paste, thrown on to a wetted filter and then washed with freshly-boiled distilled water. The clear filtered liquid should give nothing more than a slight cloudiness on the addition of a little dilute sulphuric acid. Some samples of flake-white which had been insufficiently washed contained from 2 to 11 per cent. of lead acetate removable by distilled water. In order to ascertain whether the lead carbonate and lead hydrate exist in due proportion in a sample of white lead, a weighed portion of the dry pigment, after having been dried at 212° F., should be carefully roasted in a current of dry air, and the water evolved (2 to 3 per cent.) intercepted by means of a weighed calcium chloride absorption tube. This operation, however, requires much manipulative experience, and, unless accurately performed, may lead to erroneous conclusions.

(2) Of the intentional adulterations of white lead the following are the most usual :

a. Heavy spar, that is, native barium sulphate; or the same compound artificially prepared (permanent white, blanc fixe).
b. Gypsum.
c. China-clay.
d. Whitening or chalk.
e. Lead sulphate.

The first of these adulterations is by far the most usual Barium sulphate, in the form of finely-ground barytes, or heavy spar, is the material employed on the large scale for cheapening the cost of production of ordinary white lead; precipitated, that is, artificially-prepared, barium sulphate is used in the case of the finer makes of this pigment. In either case the sophistication is very readily recognised. Pure flake-white, for example, loses 14½ per cent. of its weight when strongly heated so as to drive off its carbonic acid and water, but 'Venice' white, which is white lead and barium sulphate mixed in the proportion of equal parts, loses, under such treatment, no more than 7·3 grains per 100. 'Hamburg' white, with 33 per cent. only of white lead, loses 4·8 per cent., and 'Dutch' white, of which three-fourths are barium sulphate, gives off no more than 3·8 per cent. 'Crems,' or 'Cremnitz' white, is, or ought to be, pure white lead. The complete solubility of pure white lead in dilute nitric acid may also be made use of to detect the presence of barium sulphate, which will remain undissolved as a dense white powder. The adulterations with gypsum, china-clay, whitening and lead sulphate, can be recognised only by further tests. Gypsum, for instance, gives off water when heated, and 1 part of it dissolves in 420 parts of water. China-clay also gives off water when heated, but is insoluble in water, and only slightly soluble in nitric acid. Whitening dissolves in all mineral acids, but lead sulphate is practically insoluble. After all, the detection of barium sulphate is the only point with which the painter need concern himself. It will therefore suffice if he ascertain that a sample of white lead is of first-rate colour and body, contains no sensible quantity of lead acetate, loses when heated 14½ per cent. of its weight, and dissolves perfectly in dilute nitric acid.

It has been observed that white lead is less liable to be blackened by sulphuretted hydrogen and by other sulphides when it contains a small quantity of baryta white, or of lead sulphate thoroughly incorporated with it by grinding. This observation opens the door to adulteration, it is true: and it is perhaps wiser to rely upon the protection furnished by resinous mediums and a final coat of mastic varnish rather than upon any admixture with other white substances.

The drawbacks attendant upon the use of white lead as a paint are its poisonous character, its sickly and noxious smell when used with oil, and its liability to discolour when exposed to sulphuretted hydrogen or any sulphide soluble in water. On the other hand the quality of the whiteness of the best flake-white is unimpeachable: the paint works admirably in oil, and has great body; moreover, flake-white not only mixes perfectly and safely with the majority of permanent pigments, but it serves to impart to slow-drying colours its own strongly siccative character. Besides all these merits white lead possesses a valuable property, which has scarcely been clearly recognised or duly appreciated. For when an old oil-picture is carefully examined, it will generally be found that if any portion of its surface (of the paint, not the varnish) show decided contractions and cracks, these are precisely those portions into which white lead has entered in smallest proportion, if at all. The most translucent parts, the rich glazings and the deepest shadows may be fissured, but not the high lights: examples illustrative of this point are referred to in Chapter XXIV. of the present volume. This property of white lead seems to depend upon the combination which takes place between a part of the oil with which it is ground and a part of the lead hydrate which it contains. A kind of lead-soap (lead linoleate) is thus formed; this compound imparts a degree of toughness and elasticity to those films of oil paint into which lead-white enters to any considerable extent. A very pretty and instructive experiment proving the formation of this lead-soap is made thus. On a sheet of glass a smear of flake-white in oil is spread; then a few drops of weak sulphuric acid (10 per cent. of the real acid) are mixed thoroughly with it by means of a glass rod or glass spatula. The whole may be easily incorporated into a cream-like mixture, for the acid decomposes the lead-soap and destroys the oily or hydrofuge character of the paint. No such commixture of acid can be made with zinc white in oil or baryta white in oil, for in these paints the oil has not been saponified.

Flake-white becomes brown, grey or black when exposed to the action of sulphuretted hydrogen, ammonium sulphide, or any metallic sulphide soluble in water. This discolouration, which is due to the formation of lead sulphide, occurs more readily in the presence of moisture: it is favoured by darkness to such an extent that a piece of perforated cardboard laid upon a dry oil-painted surface of white lead will, after a few weeks' exposure, give a white pattern on a buff ground. But, after the removal of the perforated card and subsequent exposure of the painted surface to strong light, this pattern will disappear, the coloured sulphide of lead being oxidised into the white sulphate. The same change may be more speedily brought about by means of a solution of hydrogen peroxide. By laying a sheet of white filter-paper soaked in this liquid upon the discoloured lead-priming of a prepared canvas the original colour of the paint may be gradually brought back. This method is not available in the case of drawings or water-colour paintings in which flake-white has blackened; but even these may often be successfully treated by exposure to moist ozone, or by light touches of a solution of hydrogen peroxide. Old silver-point drawings having the lights heightened with lead-white may sometimes be thus restored to their pristine state.

The specific gravity of the best flake-white is 6·6; 100 parts by weight of it require from 11 to 13 parts of linseed oil in order to form an oil-paint of suitable consistence. It is sometimes ground with poppy oil when a particularly pure white product is demanded. The yellowish tint of some makes of white lead is occasionally neutralized by the addition of a trace of indigo or of artificial ultramarine. Burnt or roasted white lead is sometimes used as a pigment. It is of a cream-colour, a buff, or a pale yellowish salmon, according to the temperature at which it has been prepared, or the length of time during which it has been heated.

Lead Sulphate.—Many attempts have been made to utilize the sulphate of lead ($PbSO_4$) as a pigment. This compound, which is nearly insoluble in water and in dilute acids, is almost, if not entirely, destitute of poisonous properties owing to this insolubility, although as ordinarily prepared it possesses neither the pure whiteness nor the body of white lead. But under the name of Freeman's white lead, or non-poisonous white lead, a paint has been lately introduced which is likely to prove a formidable rival to ordinary white lead. It is essentially lead sulphate, and is prepared by precipitating lead acetate solution with sulphuric acid. But this precipitate is subjected to a special process of grinding with small quantities of zinc white and barium sulphate, and acquires thereby a considerable increase of density and opacity. Being, when ground in oil, not only destitute of the disagreeable smell of white lead, but much less readily darkened by sulphuretted hydrogen, Freeman's white possesses distinct advantages over the more common paint. It may be mixed with other permanent pigments without injuring them.

Lead Oxychloride.—Pattinson's white ($PbClHO$) does not possess any advantage, as a white pigment for artists' use, over the ordinary flake-white. Similar verdicts may be pronounced as to the eligibility of several other white compounds of lead, such as the sulphate, the antimonite, the antimoniate, and the tungstate of this metal.

The blanc d'argent of the French is supposed to be pure lead carbonate free from any hydrate, but the great majority of the specimens which I have examined are nothing but flake-white of good quality. For general use as a white pigment, both alone and in admixture, the best flake-white, with all its defects, presents distinct advantages over pure lead carbonate free from lead hydrate.

ZINC-WHITE: *Chinese-white—Blanc de Zinc—Zinkweiss.*

The substitution of carbonate of zinc for white lead seems to have been first suggested by Courtois of Dijon in 1787. After several unsuccessful attempts to introduce either the carbonate or the oxide as an oil paint, the latter began to be used about 1849-50, shortly after Leclaire had shown how to prepare an oil suitable for making the paint dry. We believe that it had been frequently employed as a water-colour many years before 1849. So early as 1834 Messrs. Winsor and Newton prepared a peculiarly dense form of this pigment under the name of Chinese-white.

For the preparation of the best zinc-white it is essential that the zinc be pure; especially should it be as free as possible from the metal cadmium. The zinc is heated to the distilling-point in crucibles or retorts set in a furnace; the vapour, meeting with air, burns into the white oxide, which condenses in a series of chambers. The contents of these chambers vary somewhat in purity of tint; the presence of some metallic zinc generally imparts a greyish hue to the zinc oxide nearest the crucibles or retorts. By selecting the densest and whitest product, and then submitting this to powerful mechanical compression when red-hot, an excellent pigment having a dense body is obtained. Zinc-white prepared in the wet way, as by the action of lime-water upon zinc chloride, is inferior in substance to that made as above described, while that obtained directly from blende is of bad colour.

As an oil-paint, zinc-white is a bad dryer. Instead of being ground in raw poppy or linseed oil, an oil rendered highly siccative by borate of manganese should be employed. In spite of its unquestionable merits, zinc-white in oil cannot be recommended as a complete substitute for flake-white. When used freely, it often shows a tendency to crack and scale, besides becoming with age more translucent, or rather, less opaque. For water-colour painting, tempera, and for fresco, zinc-white is practically perfect, being unchangeable in hue or opacity under the most adverse influences. Paper washed with zinc-white, either alone or tinted with a coloured pigment, affords a good ground for silver-point or pencil drawings. There is a peculiar 'tooth' in the zinc-white which freely brings off the silver or graphite from the pencil, and serves to fix it on the prepared surface.

The purity of zinc-white is easily tested. Heated in a tube, it should yield no volatile product, and should suffer no permanent change of hue. It should dissolve completely without effervescence in boiling dilute nitric or hydrochloric acid. If, on heating, it acquires a permanent yellowish hue, giving off moisture at the same time, white lead is probably present. If it does not dissolve completely in acid, it probably contains barium sulphate; if effervescence occurs during solution, either whitening, or white lead, or zinc carbonate is present. Zinc carbonate, however prepared, is inferior in whiteness and body to the oxide.

Zinc sulphide has been prepared as a paint; its liability to evolve sulphuretted hydrogen renders its use as an artists' pigment dangerous, for there are several other colours upon which it would exert a deleterious action. It has very considerable body.

BARYTA-WHITE: *Permanent-white—Blanc Fixe—Permanent-weiss.*

The mineral known as heavy spar, or barytes, has been used as a white paint, particularly as an adulterant for white lead. However finely it may be ground, it is always very inferior in body and covering-power to the artificially-prepared barium sulphate. To make this, a cold solution of barium chloride of specific gravity 1·19 is prepared, and to it is gradually added in the cold, and until no further precipitate is formed, dilute sulphuric acid of 1·245 specific gravity. The barium sulphate is washed with cold water until the wash-waters are entirely free from acid; for many purposes to which the product is applicable (fresco and tempera painting) it should be kept under water.

Baryta-white is absolutely unalterable by an impure atmosphere, and is without action upon other pigments. It does not work well in oil, but a mixture of flake-white and baryta-white, in the proportion of 2 to 1, presents the advantage of being very much less affected by sulphuretted hydrogen than flake-white.

The artificial baryta-white may be distinguished from the natural by its much finer state of division, by its greater body, and by the purity of its whiteness. Baryta-white is not adulterated, but its almost absolute insolubility in hydrochloric or nitric acid enables it to be at once distinguished from zinc-white or white lead.

Several mixtures of barium sulphate and zinc sulphide have been introduced as pigments; they are not suitable for the palette of the artist.

Amongst other white pigments which we need not describe are antimonious oxide, antimonious oxychloride, lead sulphite, lead tungstate, lead antimonite, and lead antimoniate. Not only are these compounds difficult to prepare in a satisfactory condition of purity and whiteness, but they are liable to turn yellow or dull in impure air.

It should be stated here that the tests described in the present chapter, and in all the other chapters on pigments can be tried only after the removal of the vehicle with which they have been ground. Oil may be removed by means of benzene or turpentine-spirit, gum by treatment with distilled water.

CHAPTER XIV

YELLOW PIGMENTS

YELLOW OCHRE: *Roman Ochre—Golden Ochre—Mineral Yellow—Brown Ochre—Oxford Ochre—Ocre jaune—Gelben Ocker.*

THE distinction between the yellow ochres and the red ochres, whether natural or artificial, depends upon a perfectly definite chemical difference. The colour of every one of these pigments is due, indeed, to iron, and to iron in the same state of oxidation; but the iron oxide in the yellow and brown ochres is chemically united to water, while in the red ochres it is nearly or quite anhydrous—that is, dry. In chemical language, then, we may say yellow ochre is a ferric hydrate, red ochre a ferric oxide. But, when we proceed to examine a number of samples of yellow ochre, we find, not merely different proportions of ferric oxide to combined water—that is, different ferric hydrates—but we find also very variable proportions of intruding or accessory constituents. In fact, yellow ochre represents not less than three mineral species, and it occurs associated with many impurities, the latter consisting mainly of silica, of clay, of rocky débris, with traces of gypsum, of iron or copper pyrites, and of humus or peaty acids. The three fundamental minerals, in order of frequency, which may be traced in various yellow ochres, are these:

Brown hæmatite, or *limonite*, consisting of two molecules of ferric oxide combined with three molecules of water, and represented by the formula $2Fe_2O_3, 3H_2O$.

Yellow hæmatite, or *xanthosiderite*, consisting of one molecule of ferric oxide combined with one molecule of water, and represented by the formula Fe_2O_3, H_2O.

Bog-iron ore, or *lymnite*, consisting of one molecule of ferric oxide and three molecules of water, and represented by the formula $Fe_2O_3, 3H_2O$.

It is probable that all the numerous varieties of yellow ochre, from the countless localities of this substance, belong essentially to one or other of the above species of iron minerals, although the frequent presence of such impurities or accessories as silica, iron silicates, and clay renders the identification very difficult.

An analysis of a fine sample of yellow ochre, from the pit on Shotover Hill near Oxford, gave the following percentages:

Hygroscopic moisture	7·1
Combined water	9·0
Ferric oxide	13·2
Alumina	6·3
Silica	61·5
Calcium sulphate	1·4
Undetermined	1·5

The varying hues of yellow ochres depend mainly upon two differences of composition. One of these is the amount of white clay or of silica present in them—this lightens the colour; the other is the presence of ferric oxide, which gives them a ruddier or warmer hue. All, when burnt—that is, calcined—lose their essential water, and become converted into various kinds of red ochre, light red, etc. The varieties which contain much silica and clay (ingredients which, even in good yellow ochres, often amount to two-thirds of their weight) yield the less translucent and paler tints of some of the burnt red ochres. India furnishes a great variety of hues of yellow ochre, but our chief supplies come from France, Italy, Germany, and Spain. Quite recently excellent ochres have been obtained from the district of Dubbo in New South Wales. Some of the English ochres (from Oxfordshire, Derbyshire, etc.) are of fine quality.

Yellow ochre is generally prepared for use as a pigment first of all by careful selection of the best pieces, and then by the familiar process of elutriation, or washing over. Thus it is at once freed from sand or other coarse particles, and from any soluble salts which it may contain. Immediately before being ground in oil, it should, however, be dried at a temperature a little below that of boiling water, as it is liable to contain hygroscopic moisture in addition to its necessary constitutional water.

Yellow ochre is one of the most ancient pigments, having been used by the Egyptians, the Greeks, and the Romans. It is the oichra of Theophrastus. Pots of yellow ochre were found at Pompeii. It has stood, with very little change, the test of centuries. It certainly does become, in all media, slightly darker and warmer in hue after prolonged exposure to light. The change, however, is slight; moreover, it soon comes to a stop. It is probably due in part to a slight loss of constitutional water from the ferric hydrate, and in part to increased translucency. Yellow ochre, so long as it is exposed to air and light, is not darkened by sulphuretted hydrogen. It is without action on other pigments, although the statement has often been made, on quite insufficient grounds, that paints which are damaged by contact with *metallic* iron are likewise damaged by yellow ochre and by the red oxide of iron. For instance, true Naples yellow is undoubtedly spoilt by contact with a steel spatula, because the metal of the latter takes away oxygen from, or 'reduces,' the lead antimoniate of which the former consists. But such an action is impossible with yellow ochre, for this iron compound is a stable substance, containing already all the oxygen it can take up. It is possible, notwithstanding, that ochre may injure the hue of some lakes, such as yellow lake and crimson lake, by replacing in part some of the alumina with which the colouring matter is united. But as such lakes are worthless, from their extreme instability when exposed to light, when used alone, such probable action of ochre upon them need scarcely be considered. With reference to the darkening or embrowning of yellow ochre in oil, this change is in part due to the change of the oil, of which this paint usually contains as much as 43 per cent., and in part to an increase in the translucency of the mixed paint.

Yellow ochre is little subject to adulteration, for it is too cheap a pigment to make it worth while to substitute other substances for it. But sometimes the golden and richer-coloured varieties have been found to have had their colour enhanced by the addition of turmeric, or of certain other fugitive or semi-permanent yellows of organic origin. The majority of such additions may be detected by pouring a little liquor ammoniæ mixed with spirits of wine upon some of the ochre placed on a filter paper in a funnel: the liquid passing through will be colourless if the ochre be genuine. An ochre which when heated in a test-tube gives off, besides water, fumes which partially condense into a coloured or tarry matter on the glass, contains organic matter, naturally present or artificially added, and is generally of inferior permanence. An artificial yellow ochre is made by acting upon solutions of iron salts with metallic zinc, and thoroughly washing the precipitate obtained.

Brown ochre is an approximately pure limonite: raw sienna is very nearly related to it (see further on), but cologne earth, raw umber, caledonian brown and vandyke brown are distinct substances. An artificial brown ochre is prepared by heating yellow ochre with 4 per cent. of common salt to a low red heat.

Under the name of cyprusite a peculiarly bright lemon-coloured earth has been imported from Cyprus as a pigment: it consists essentially of a hydrated ferric sulphate: it is not likely to prove a safe pigment for artistic use.

CADMIUM YELLOW: *Orient Yellow—Aurora Yellow—Orange Cadmium—Sulphide of Cadmium—Jaune Brillant—Jaune de Cadmium—Cadmium Gelb.*

The metal cadmium, which is nearly related to zinc both chemically and physically, was discovered by Stromeyer in the year 1817. To one compound only of cadmium, the sulphide, are due all the hues and tints from the palest lemon cadmium to the fiery orange-red. This compound is represented by the formula CdS, and contains 112 parts by weight of cadmium to 32 parts of sulphur. As commonly prepared, cadmium yellow is of an orange hue; when this compound separates slowly from a solution, or is made in any way to take a dense or aggregated form, it becomes of a decided reddish orange. The orange-yellow variety, when very finely ground, becomes less red and more inclined to yellow. Some of the palest cadmium yellows contain white pigments, or flour of sulphur, added to reduce their depth of colour: the presence of free sulphur is sufficient to make any pigment ineligible.

There are two well-known processes for making cadmium

yellow. In one of these pure cadmium oxide is heated in a covered crucible with pure sulphur in excess. In the other process, which yields pigments of greater brilliancy and beauty, a soluble salt of cadmium, such as the chloride or sulphate, is precipitated in the presence of a little free acid, by means of a solution of sodium sulphide, or preferably, of a stream of sulphuretted hydrogen. The hue of the product inclines to red when the solution is strong, hot and faintly acid; to yellow when it is weak, cold, and neutral. It is necessary to state that all the materials used must be pure. The presence of a little zinc is immaterial; but iron, lead, bismuth, and any metal giving a coloured sulphide, even in traces, are seriously detrimental to the beauty of the product. The precipitate of cadmium sulphide, after having been thoroughly washed with boiling distilled water until the wash-waters no longer redden blue litmus paper, is collected on filter-papers and dried in the water-oven. In order to remove any free sulphur that may be present the dry cadmium yellow may now be digested in a suitable vessel with pure carbon disulphide. After this treatment the pigment is once more dried, and is then ready for grinding in oil or other vehicle. Cadmium yellow, prepared by the process last described, presents a satisfactory degree of permanence, and has no action on white lead when both pigments are ground together in oil. But a curious change is noticeable when the orange-red variety of this pigment, ground in oil, is kept some time in the ordinary metallic collapsible tubes. The interior surface of the tube becomes darkened, sometimes almost black, from the formation of lead sulphide. It is certainly strange that a similar action does not occur between white lead and these deep cadmiums. For I found that the same sample of cadmium-red in oil which had blackened the metallic tube, when

9—2

some of it was laid upon flake-white in oil, and kept for years, had not darkened the lead compound anywhere, even at the surface of contact. Moreover, cadmium yellows mixed with flake-white prevent, as do many other substances, such as baryta white, lead sulphate, etc., the ready darkening of this lead paint by sulphuretted hydrogen. On the other hand, the cadmium yellows act with great energy upon some of the pigments containing heavy metals. Emerald green, for example, is rapidly ruined by cadmium sulphide, both in water and in oil; cadmium yellow and emerald green (Schweinfurt green) are absolutely incompatible. Chrome yellow and true Naples yellow are also darkened by admixture with cadmium yellow, at least after a time.

While the stability of what may be called the normal cadmium yellow or orange is pretty well assured, both as an oil and a water colour, a very different verdict must be pronounced upon pale and lemon cadmium when used in water-colour painting. When thus used these pigments do not merely fade, but acquire a somewhat greyish hue. The following observations throw some light upon these changes. During the year 1876 I prepared a number of samples of cadmium yellow and orange. All were obtained by the action of sulphuretted hydrogen upon solutions of cadmium chloride. The products ranged in hue from a lemon colour to a deep orange, according to the strength of the solution, the presence or absence of free acid, and the temperature at which the precipitation of the pigment took place. After due washing and drying the various samples were put into bottles and preserved in my laboratory. They were never exposed to direct sunshine. On examining them from time to time it was noticed that the specimens of medium depth, having a yellowish orange hue, kept their hue perfectly, while the orange-red varieties exhibited a curious phenomenon of alteration. The loose friable lumps into which the powder had aggregated were distinctly paler on the outside than in the interior, while the parts of the contents of the bottles which had been most exposed to light were paler than those which had been comparatively shaded. But a still more marked change had taken place in the samples to which the term 'pale' cadmium might be applied. These had generally become still paler, almost straw-coloured, especially where most exposed to light; but in some of the specimens orange specks were observed, resembling in hue what is usually called 'middle' cadmium. From the above observations it would seem that there is a tendency in differently tinted cadmium yellows to return to what we may call the normal or medium hue, but that the palest varieties are most subject to change. This change seems to arise in part from oxidation and hydration, for the bleached specimens gave indications of containing some white cadmium hydrate, when heated giving off a little water, and becoming brownish from the formation of the brown oxide of cadmium. Such a bleaching of pale cadmium, if my explanation be correct, is in a measure explicable if we recollect that this variety occurs in a very fine state of division, and on this account is more liable to chemical change. In water-colour painting, where there is no effective protection through the presence of a hydrofuge medium, this fading of pale cadmium is notorious. In oils this cadmium, like the others, is practically permanent. I regard the passage of the pale and of the deep cadmium yellow when in powder into the normal or middle variety as dependent chiefly, if not entirely, upon a molecular change. Moreover, the pale cadmiums are rarely found free from admixture, and their alterability may be in part owing to the foreign ingredients they contain. The recent researches of G. Buchner and N. von Klobukoff confirm the conclusions drawn from my early experiments. There can be no doubt that cadmium sulphide exists in two if not in three molecular states, differing not only in colour but in crystalline form and in specific gravity. Thus pale cadmium has the specific gravity 3·9, while the red modification is denser—4·5. And when the pale variety, dry and in powder, is rubbed strongly with a piece of agate, its colour deepens and reddens in a very decisive manner. This change is clearly analogous with that suffered when the yellow mercuric iodide is altered into the scarlet form by pressure. It is perhaps safer to employ an ivory palette knife rather than one of steel in manipulating the cadmium pigments.

Cadmium yellows are sometimes adulterated with Indian yellow, baryta and strontia chromates, and chromates of lead. Indian yellow shows its presence by blackening and giving off tarry fumes when the pigment, in the state of dry powder, is strongly heated in a test-tube. The chromates may be detected by the green colour produced when the sample is warmed with alcohol and dilute sulphuric acid. The lead chromates or chrome-yellows, and the orange and red basic chromates of the same metal will blacken when the substance in which they are present is moistened with weak ammonium sulphide. Free sulphur in pale cadmium yellows comes off as a vapour when the sample is heated, but it may be better detected by the solvent action upon it of carbon bisulphide. Baryta white may be detected by its insolubility in hot strong hydrochloric acid, in which cadmium sulphide dissolves.

Cadmium red and cadmium orange are slightly translucent when compared with the paler and yellower varieties of this pigment, and possess very full and glowing hues. They work well as oil and water colours. Mixed with zinc-white or flake-white, deep and middle cadmiums yield several beautiful colours, some of which closely resemble the different varieties of true Naples yellow, and are now employed very largely in lieu of the latter pigment. Pure cadmium yellow, when heated moderately, becomes orange-red or red, but regains its pristine hue on cooling. If, however, the heat be considerably raised in the presence of air, some of the sulphur in the compound burns, and the residual mass presents a dull brown colour. 'Manganese oil' accelerates the drying of the cadmium colours, which is sometimes inconveniently slow.

As the metal cadmium can be purchased for five shillings a pound, and as this amount will yield twenty ounces of the dry pigment, cadmium yellow should be obtainable at a moderate price.

AUREOLIN: *Cobalt Yellow—Jaune de Cobalt—Kobaltgelb.*

Origin and Composition.—This remarkable artificial yellow pigment was discovered by Fischer. It is a compound of the nitrites of cobalt and potassium. Usually it is free from water, but it sometimes contains three molecules, and is then represented by the formula $K_6Co_2(NO_2)_{12}$, $3H_2O$. Other proportions of water also occur; but when the compound contains four molecules, its hue is somewhat greenish. The dry or anhydrous variety is best made by mixing a solution of a cobaltous salt, strongly acidified with acetic acid, with a concentrated solution of potassium nitrite, and keeping the mixture warm. Perhaps a pigment of finer hue is obtained by passing a stream of nitric oxide gas mixed with air into a solution containing nitrate of cobalt and a little acetate of potassium; from time to time a little potassium carbonate is added. It is not much affected by caustic potash solution or by hydrochloric or nitric acid. It is very slowly attacked and blackened by solution of sulphuretted hydrogen, but is at once destroyed by ammonium sulphide. It is slightly soluble in cold water.

Aureolin is of a pure yellow colour, and is almost transparent whether used in water or oil painting. In oil some samples dry with great difficulty, and become very dirty if exposed to the air during the progress of desiccation; other samples dry and harden too quickly—the exact cause of this difference of deportment has not been ascertained. Moreover, as ordinarily ground in oil, it requires a very large proportion of the latter medium—nearly 2 parts (by weight) for 1 part of aureolin powder. These defects may be easily remedied by heating the slow-drying variety of the ground pigment to 212° F. immediately before the addition of the oil, and by using, instead of raw linseed oil, the siccative linseed oil, prepared by means of borate of manganese. Thus prepared, aureolin not only dries quickly, but it retains its purity of hue; moreover, a surface of the dried oil pigment will yield nothing to a wet cloth passed over it instead of staining it yellow. The reason why this staining occurs with ordinary aureolin ground in oil, even when it has at last become dry, is that the oil, though present to the extent of 67 per cent., does not suffice to protect the particles of pigment from the solvent action of moisture. The quick-drying variety of aureolin should be ground in poppy oil.

Aureolin properly prepared in oil, as described above, does not fade by exposure to sunlight, nor does it darken, except so far as the admixed oil is concerned. As a water-colour aureolin is practically permanent, even in sunlight, as the following figures show:

Original intensity	10
After two and five years	10
After ten years	9

The fading of fugacious organic pigments, such as the lakes from cochineal, is accelerated by their commixture with aureolin, which particularly hastens the destruction of indigo, even in oil. The aureolin cannot so act without being itself likewise affected; it generally becomes, under such circumstances, of a brownish hue.

It is easy to learn whether a sample of aureolin is free from combined water by heating a small portion somewhat strongly in a long test-tube; dew will condense upon the upper part of the tube if water be present in the pigment. The presence of yellow organic matters in imitative or adulterated aureolins may generally be detected by mixing some powder of the suspected sample with spirits of wine and a few drops of strong ammonia; the liquid becomes red, orange, or yellow if the aureolin be not pure. Aureolin containing chrome yellow is blackened by a solution of sulphuretted hydrogen.

LEMON YELLOW: *Baryta Yellow—Barium Chromate—Yellow Ultramarine—Permanent Yellow—Jaune d'Outremer.*

Of all the chromates which have been used in painting, barium chromate is the most stable. It has a pure yellow colour, with a not inconsiderable degree of opacity. It works smoothly.

Lemon yellow is often made by mixing solutions of neutral potassium chromate and of barium chloride, both liquids having been previously heated to 100° C. A still better plan is to take equivalent proportions—namely, $25\frac{3}{4}$

parts by weight of pure crystals of barium chloride and $21\frac{1}{2}$ parts of pure crystals of neutral potassium chromate—of these two compounds, and to grind them together to very fine powder. Continue the grinding, and then add gradually sufficient pure water to convert the mixture into a thin paste. The paste is then heated to 100° for fifteen minutes, thrown on a filter, washed with abundance of pure water, dried, and ground.

Properly prepared lemon yellow may be mixed with most other stable pigments without suffering change. It is not blackened like the lead chromes by sulphuretted hydrogen, but it has a tendency, as a water-colour, to become greenish when long exposed to this gas or to impure air. In oils it is very useful, for although some organic pigments may give it a greenish cast by reducing it in part to green chromic oxide, yet it may be safely associated with aureolin, with madder carmine, and with Prussian blue. Lemon yellow may be used in fresco.

Strontium chromate is very often—we may say, generally,—substituted for true lemon yellow, but it is less stable, and has the further defect (for water-colour work) of being decidedly soluble even in cold water, so that light washes of it may be found to sink into the paper and to partially disappear. The most common adulteration of lemon yellow is with pale chrome; of course, sulphuretted hydrogen detects this falsification by darkening or blackening the pigment. Strontium chromate is distinguished from barium chromate by its dissolving in boiling water to such an extent as to yield a solution having a strong yellow colour. It may be prepared in the same way as the chromate of barium. Zinc chromate and calcium chromate are yellow pigments of inferior value.

GAMBOGE : *Gomme-gutte.*

Origin.—This gum-resin is produced by several species of Garcinia. Siam gamboge comes from *G. Hanburyi* (Hook. f.); Ceylon gamboge from *G. Morella* (Desv.). There are other species from which the same product is obtained in various parts of India, as *G. Cambogia* (Desrouss.) and *G. elliptica.* The fine, deep-coloured gamboge, produced by the Burmese *G. heterandra* (Wall.), may prove to be superior to Siam gamboge, but it has not yet become an article of European trade. Gamboge is a mixture of a gum soluble in water, and a resin which is soluble in alcohol, chloroform, ether, etc. The pipe-gamboge of Siam, which is as pure as any variety met with in commerce, contains about 78 per cent. of resin, and 18 of gum. The resin, which is the true colouring-matter, may be easily obtained pure by crushing pipe-gamboge into fine powder, mixing it with a little water, and then shaking up the mixture with ether; the ether dissolves the resin alone. From the ethereal solution the colouring-resin is recoverable by evaporation; but it is better to add a little drying-oil and some copal-varnish before driving off the ether by means of a very gentle heat. The coloured, semi-fluid mass which then remains may be preserved in bottles or tubes for use as an oil-paint. The resin of gamboge has the properties of an acid, and forms yellow, orange, or brown compounds, with soda, lime, baryta, and other bases. Some of these compounds might prove useful as paints.

Gamboge was used by the early Flemish oil-painters. In the seventeenth century it was largely employed to give a golden hue to the embossed leathers for which Amsterdam was famous.

In water-colour painting gamboge is not trustworthy. It

is unaffected by sulphur compounds, but is darkened by ammoniacal fumes, and slowly bleached by strong heat. Some samples prove, however, far less fugitive than others. In two years' exposure to sunlight, one sample of cake-gamboge lost more than half its original intensity; while a sample of moist gamboge, bought at the same time from the same maker, retained nine-tenths. The same sample of moist gamboge, after seven years, still showed seven degrees out of the original ten of intensity.

As an oil-colour, gamboge affords a rich, transparent, golden, amber hue; it has some claims to the consideration of artists. To secure its permanence, admixture with oil alone does not, however, suffice; a resin such as copal, or wax, or paraffin, must be used also. Some of Sir Joshua Reynolds' trials of gamboge prove this, those with oil alone being a name only now; while those with resin, or wax, retain their original hue very fairly, though they were spread upon the canvas in 1772. It must, therefore, be remembered that reliance cannot be placed upon the permanence of the ordinary gamboge oil-paint as met with in commerce.

Gamboge, from its resinous nature, shows, when laid on thickly as a water-colour, a rather shining surface. It appears to have little or no chemical action on other pigments (with the exception, perhaps, of white lead), although, if it be mixed with anything which contains lime, or other alkaline compounds, it becomes brownish, and darkens. Gamboge forms beautifully clear and rich greens with Prussian blue or indigo, but its place in water-colour painting may be advantageously taken by aureolin and even by Indian yellow. When mixed with baryta yellow or cadmium yellow, the permanency of gamboge is enhanced.

INDIAN YELLOW: *Piuri, Purree, Peori—Jaune Indien—Indisch-gelb.*

This remarkable pigment is obtained at Monghyr, a town in Bengal, from the urine of cows which have been fed upon mango-leaves. It generally occurs in the bazaars of the Panjáb in the form of large balls, having an offensive urinous odour.

Indian yellow is an impure magnesium salt of euxanthic acid. The essential part of it is a compound containing 4·5 per cent. magnesia, 18·7 per cent. water, and 78·7 per cent. euxanthic anhydride; but this substance is always associated, even in the most carefully purified samples of prepared Indian yellow, with various impurities both mineral and organic. The pure magnesium euxanthate is represented by the formula $C_{19}H_{16}MgO_{11},5H_2O$.

For artistic purposes the crude imported Indian yellow is thoroughly powdered, and then washed with boiling water, until the liquid filtered from it is no longer coloured; a brown impurity, and much of the evil smell, is thus removed. The colour of the washed product is enriched by leaving it in contact for a day or two with a saturated solution of sal-ammoniac, and then repeating the treatment with hot water.

Thus purified, this pigment presents a translucent orange-yellow colour of great depth and beauty. Ground in oil, some specimens are practically unchanged, even after long exposure to sunlight, any darkening they show being due either to imperfect purification, or to the change of the associated oil. Such change is reduced to a minimum if poppy oil be substituted for linseed oil, or if the latter be previously treated with manganese borate. On the other hand, I have met with specimens of Indian yellow ground in oil which, after five years' exposure, have lost nearly one-third of their original depth, and have, at the same time, become rather reddish-brown in hue. As a water-colour, Indian yellow retains its hue unimpaired when exposed to diffused daylight; sunlight very slowly bleaches it, the hue it acquires being somewhat brownish. The rate of alteration and of reduction in force caused by sunlight may be approximately represented by these figures:

Original intensity	10
After 2 years	9
After 5 years	7
After 7 years	6
After 10 years	5

As a general rule, Indian yellow suffers no change by admixture with any pigment itself permanent, nor is it affected by sulphur compounds. True Naples yellow, however, most of the chromates, and probably aureolin also, tend to embrown it to some extent.

Indian yellow which has been adulterated with lead chromate (chrome yellow) becomes dark-brown when moistened with ammonium sulphide.

A fine yellow pigment may be prepared from the euxanthic acid, which is the characteristic constituent of Indian yellow, by throwing it down in combination with the two bases—alumina and magnesia. The following directions may be followed: Dissolve 1 part of pure euxanthic acid in just sufficient dilute ammonia. Pour the solution into a liquid prepared by dissolving 45 parts of potash-alum, 15 parts Epsom salts, and 6 parts sal-ammoniac in 250 parts of water. Now cautiously add dilute ammonia to the mixture, stirring all the time, and avoiding any excess of ammonia. The precipitated pigment is to be thoroughly washed, and then pressed, dried, and ground.

MARS YELLOW: *Mars Orange—Artificial Ochre—Jaune de Mars.*

This pigment is a kind of yellow ochre prepared artificially. It may be made by precipitating a salt of iron mixed with alum by means of caustic soda, or potash, or lime. The salts of iron used are either green vitriol (or ferrous sulphate) or the ferric chloride. If green vitriol be employed the precipitate formed gradually becomes yellow on exposure to the air. Upon the proportion of alum mixed with the iron salt depends the depth of the yellow colour in the product, for the alumina precipitated with the iron hydrate acts as a diluent of the colour. When lime is used as a precipitant for the iron compound (if this be green vitriol or ferric sulphate) calcium sulphate, that is, gypsum, comes down along with the ferric hydrate and basic ferric sulphate, and serves to lighten the colour.

By submitting the different varieties of Mars yellow to various degrees of heat, with or without a little nitre, a number of products of different hues are obtained, including Mars orange, Mars red, Mars brown, and Mars violet. All these preparations require very thorough washing to fit them for use on the palette of the artist.

The Mars colours are permanent when carefully prepared, and thoroughly purified from soluble salts. They seem sometimes to have a slightly injurious effect upon a few of the best semi-permanent pigments of organic origin, such as the madder colours. This action may be due to the ferric hydrate in them combining with the colouring matter, and displacing some of the alumina previously united with it. In this direction it is probable that Mars yellow will be more active than the deeper-coloured pigments produced by calcining it at various temperatures.

NAPLES YELLOW: *Jaune de Naples—Jaune d'Antimoine—Neapel-gelb—Giallo di Napoli.*

Under this name three different substances are included. The pigment generally sold in England as 'Naples yellow' is an excellent imitation made by mixing cadmium yellow or deep cadmium with a white, preferably a zinc white. But a true Naples yellow, which is a basic lead antimoniate, is still procurable from some artists' colourmen.* This preparation is sometimes made by heating together for two hours a mixture of 1 part tartar emetic, 2 parts nitrate of lead, and 4 parts common salt, all the ingredients being of the purest quality, and the heat not exceeding that at which common salt fuses. A more recent process, in which zinc oxide is introduced among the materials which are heated together, yields a paler but excellent product. A bright pale variety of yellow ochre seems to have formerly gone under the name of Naples yellow.

This antimonial yellow has been known from very early times as an enamel colour. It has been found upon Babylonian bricks at least 2,500 years old. Persian pottery as early as the thirteenth century of our era is occasionally decorated with antimonial yellow.

In oil the genuine and the imitative Naples yellows are quite permanent, so far as light is concerned, but the genuine kind is liable to be darkened, like other lead compounds, by air containing sulphuretted hydrogen. In water-colour painting genuine Naples yellow is quite inadmissible, for it blackens rapidly, but irregularly, in the presence of mere traces of sulphur compounds. This

* Roberson; Edouard, of Paris; Schönfeld, of Düsseldorf; and others.

blackening, like that of lead white under similar conditions, is much more marked in darkness than in light.

Naples yellow, in contact with metallic iron, tin, pewter, zinc, and several other metals, is discoloured and blackened. An ivory instead of a steel spatula, or palette knife, should be used with this pigment. The darkening in question is due in part to attrition, owing to the extreme hardness of the particles of the lead antimoniate, however finely the material may have been ground, and partly to the *reducing* effect of the above-named metals upon this antimoniate. Iron in the form of its oxide or hydrate (as in light red or yellow ochre), or in complex combinations (such as Prussian blue), does not exert any effect upon Naples yellow. A statement to the contrary effect has crept into a large number of technical manuals, but I have been unable to discover the slightest experimental evidence in favour of such a view. Naples yellow, however, is injured by and does injure some of the organic pigments, such as the cochineal reds and the numerous yellow lakes. But as Naples yellow cannot be used as a water colour, and as the above-named organic pigments ought to be entirely excluded from the palettes of all artists, the action in question is of little importance. Naples yellow acts upon indigo also. Indigo, however, is a pigment, to which a very high degree of permanence cannot be assigned; there is, moreover, no reason why it should be associated with Naples yellow, as other yellow pigments may be safely used to modify its hue.

Another pigment also is sold as jaune d'antimoine. It is a mixture of the oxychlorides of bismuth and lead with lead antimoniate. When carefully prepared it yields a rich paint of good body, but its use cannot be recommended to artists.

YELLOW LAKE: *Brown Pink—Citrine Lake—Yellow Madder—Italian Pink—Quercitron Lake.*

Origin.—The sources of yellow lake are numerous, but the best kind is obtained from quercitron bark from *Quercus tinctoria*, *Qu. nigra*, and *Qu. citrina*, three species of North American oak. A hot-water decoction is made, and this is precipitated by a solution of alum and dilute ammonia. A richer yellow pigment is obtained by extracting the powdered bark and alburnum with boiling dilute sulphuric acid instead of with water. The original colouring matter of the bark (quercitrin) is thus changed into a more stable compound known as quercetin. The former substance is a glucoside, the latter has the character of an acid; both may be converted into lakes by bringing them into contact with precipitating or precipitated hydrate of alumina. Yellow lake was formerly made from the fruits of various species of buckthorn, known as Persian, Turkish, or Avignon berries. The species yielding these fruits are *Rhamnus infectorius*, *R. oleoides*, *R. saxatilis*, *R. amygdalinus*, *R. catharticus*. The bark of *R. frangula* and of *R. catharticus* also yields a yellow pigment. 'Stil de grain,' and several of the continental yellow lakes, are made from the above-named berry.

Italian pink, Dutch pink, and deep yellow madder are names usually given to the richer yellow lakes of quercitron, although some of these pigments are occasionally prepared from Turkish or Avignon berries.

Beautiful and useful as many yellow lakes undoubtedly are, they should be rigorously excluded from the artist's palette. In oil most of them are very bad driers, as well as fugitive: in water-colour they generally lose nine-tenths of their colour within two years of exposure to sunlight: the residual stain is ultimately of a bluish-grey.

The following observations as to the behaviour of several members of this group, on exposure for two years to sunlight, apply to the colours as ground in oil, and as mixed with flake-white in tint:

Name	*Original Colour*	*Acquired Colour*	*Original Intensity* = 10 — *Residual Intensity.*
Laque Robert—hell-gelb	Lemon yellow	Pale straw	2
Laque Robert—dunkel-gelb	Deep lemon	Stone	4
Laque brun-jaune	Salmon	Pale rose	7
Laque brun-foncé	Yellowish-grey	Smoke-grey	8
Pale yellow madder	Pale orange	Pale buff	7
Deep yellow madder	Greyish salmon	Pale greyish pink	6

The same pigments used as glazing colours over flake-white have faded to about the same extent, but their change of hue is, in one or two cases, rather less marked.

The so-called brown pink is usually a deep quercitron lake, although it was formerly made from the berries of one of the kinds of buckthorn (Rhamnus) previously named. I have never met with a specimen of it which would stand a year's exposure to sunlight without suffering almost complete change or loss of colour both in water and in oil. And it further presents the awkward effect of becoming ultimately of a cool bluish-grey hue, a change particularly unfortunate when it has been freely used to represent foreground vegetation, or the golden lights on the near foliage of trees. Yet I am bound to confess that in Mr. W. Simpson's fifteen years' trial of certain water-colour pigments, the brown-pink has suffered comparatively little alteration. Had a portion of the original cake-colour employed been preserved for examination it might have been possible to have discovered the cause of this anomalous behaviour of the particular specimen in question.

CHROME YELLOW: *Chrome—Chromate of Lead—Jaune de Chrome—Chromgelb.*

This pigment, when of a pure yellow hue, is the neutral lead chromate. By associating it with an additional quantity of lead oxide it may be obtained of various orange and reddish orange hues. It may be made by the mutual action of a soluble lead salt, such as the acetate or nitrate, and the chromate or bichromate of potassium. Or white lead in fine powder (2 kilos.) may be boiled with a solution of bichromate of potassium ($\frac{1}{2}$ kilo.) in water (10 litres). Alum and baryta-white, or lead sulphate, are also employed in the preparation of some of the paler chrome yellows. Orange chrome and chrome red are prepared from a mixture of lead acetate ($6\frac{3}{4}$ kilos.), litharge ($5\frac{1}{2}$ kilos.), neutral potassium chromate (6 kilos.), caustic potash being sometimes used in addition. Chrome red may be obtained also by the direct action of caustic soda in solution upon the yellow lead chromate: its chemical formula is $PbCrO_4$, PbO, or Pb_2CrO_5.

The chromates of lead are peculiarly liable to change, and are quite unfitted for use in tempera or water-colour painting. In oil, especially if protected by varnish, or locked up in a resinous vehicle, these pigments show a certain measure of permanence, except when they are mingled with paints of organic origin. In fact there are two causes which militate against the integrity of the lead chromates. One of these is the tendency which they possess towards *reduction*, that is, the loss of oxygen by their chromic constituent, by which the green or lower oxide of chromium is formed. This change is brought about by many kinds of organic matter, notably by such animal or vegetable pigments as are themselves prone to oxidation. The other cause of deterioration is the presence of certain sulphur compounds which act upon the lead chromates in the same way as they act upon white lead, producing lead sulphide of a dark brown, or a grey colour.

As the artist had better avoid the employment of these pigments, further details concerning their characteristics need not be given.

Vanadium Yellow.

It has been proposed to employ the beautiful golden-bronze crystals of meta-vanadic acid as a pigment. They possess, when finely ground, an intense colour, like that of a very rich golden ochre, but less earthy, and more brilliant. This pigment has remarkable covering power, and works admirably both as an oil and a water-colour. Although the material is somewhat costly, the price for which it could be prepared need not preclude its use. But, unfortunately, this colour is not permanent. A few hours' exposure to sunshine of a water-colour wash of vanadium yellow suffices to change and deteriorate its hue in a marked degree.

King's Yellow: *Orpiment—Jaune Royal—Königsgelb.*

The yellow arsenic sulphide, though extremely beautiful in hue, cannot be relied on as a pigment. Even in oil or varnish its colour fades: Sir Joshua Reynolds' experimental canvas shows some pale brown patches which have once been king's yellow, but which now have almost entirely disappeared. Strange to say, in one of his trials, a few quite visible crystals of orpiment are preserved. As it cannot be imagined that he used this pigment in this exceedingly coarse form, it would seem that a molecular aggregation of a part of the orpiment has taken place in the lapse of years. If this change has not occurred, then we may conclude that only the largest particles of the king's yellow have escaped alteration. Under any circumstances the inadmissibility of king's yellow to the palette of the artist is obvious: moreover it cannot be safely mixed with any pigment containing lead or copper.

PURE ORANGE: *Marigold.*

Under the above names a pigment of great richness and beauty has been recently introduced. It is a kind of lake, and consists of one of the coal-tar colours known to chemists as β-nitro-alizarin thrown down upon an aluminous base. Nitro-alizarin, as its name implies, is a derivative of alizarin, one of the least changeable of all organic pigments, and the chief tinctorial product of madder. There is no doubt about the beautiful hue of this paint, a deep brownish gold: the trials already made as to its permanence in oil promise well, but it is too early to pronounce a definite judgment on this point.

CHAPTER XV

RED PIGMENTS

VERMILION: *Cinnabar—Vermillon—Zinnober.*

THE mineral cinnabar, or mercuric sulphide, occurs in many parts of Europe, and abundantly in China, and is extensively worked in New Almaden in California; it would be tedious to recount the numerous localities in which it has been, or is, found. Its colour in the mass varies from cochineal-red and red-brown to lead-grey; its powder is usually scarlet, or red. Its hardness lies between that of gypsum and that of calc-spar. It seldom contains even 1 part in 100 of impurities, but consists in 100 parts of very nearly 14 parts of sulphur by weight, united with 86 of mercury, or 1 atom of each element. The density of native vermilion is very nearly 9. Vermilion was formerly known as vermiculus, cinnabaris, cenobrium, and minium; the last name is now appropriated to red lead. Vermilion and vermiculus are derived from the Latin *vermes*, a name originally designating the 'kermes' insect found on the ilex or evergreen oak, which is still used for the preparation of a red dye. From kermes, in its turn, the words crimson and carmine are derived. The name cinnabar is supposed to be of Indian origin, and was used sometimes to designate dragon's blood, a red resin. Theophrastus informs us that two kinds of cinnabar were known to the Greeks. One of these was undoubtedly real cinnabar (chiefly from Spain), the other was red lead. Pliny's 'cinnabar' or 'minium' was true vermilion, so was the 'minium' of Vitruvius. Theophilus calls it 'cenobrium.'

One of the most curious facts concerning vermilion is that it is identical in the nature and proportion of its two constituent elements with an artificial black substance, 'Æthiop's mineral.' The red substance may be changed into the black, and *vice versâ*, and this without any loss or gain, or any alteration of chemical composition, the change being a physical or molecular one merely. The black substance is amorphous, the red crystalline.

The pigment vermilion may be made by simply grinding selected pieces of native cinnabar, or it may be obtained artificially by combining the two elements sulphur and mercury.

All the methods of preparing vermilion artificially may be grouped under two divisions. The first of these is the *dry way*, the other the *wet way*. In the former method metallic mercury 42 parts, and sulphur 8 parts, are intimately mixed and agitated together in revolving drums until they have combined. The brownish-black powder thus obtained is then submitted to sublimation in vertical iron cylinders, surmounted by heads which are connected with receivers. On sufficient heating, the mercuric sulphide sublimes as cinnabar or vermilion, the best part condensing in the retort-heads. The rest of the sublimed product (which has travelled farther) contains free sulphur, and is of inferior colour. The selected portions are next ground, moistened with water, warmed with a little caustic potash solution or nitric acid, and then thoroughly washed with boiling water. In another dry process the mercury is gradually added to the proper proportion of melted sulphur in an iron basin. When the combination (which is accompanied by a violent evolution of light and heat) is complete, the fused blackish mass is poured out, broken into fragments, heated until excess of sulphur has been driven off, and then sublimed in the way already described. Some makers add to the crude sulphide, previous to sublimation, 1 per cent. of antimony sulphide, with the object of improving the colour; the product is afterwards ground, digested with liver of sulphur, and then washed with hydrochloric acid.

There are numberless processes for preparing vermilion by the wet way. One of the best of these consists in grinding, in the presence of water, 100 parts of mercury with 38 parts of flowers of sulphur until these elements have united. The black product is then triturated at 45° C. for many hours with a solution of 25 parts of caustic potash in 150 parts of water. When the product has attained its maximum of redness and beauty, it is thrown into water, and thoroughly washed by decantation. In a second process mercury, sulphur, and pentasulphide of potassium are boiled together for three or four hours, and then the mixture is kept at a temperature of 50° C. for several days. Vermilion may also be prepared from the black sulphide obtained by precipitating a mercuric salt with a soluble sulphide, from 'white precipitate,' and from metallic mercury itself, by warming any one of these substances with a solution of an alkaline pentasulphide, and then purifying the product by means of a potash-solution heated to 45° C. It has also been found that vermilion is produced when a mixture of mercurous chloride (calomel) and sulphate of zinc is heated to 45°—50° C. with an excess of a solution of sodium thiosulphate.

Except where carmine or realgar (red sulphide of arsenic) is present, a very simple test suffices to ascertain whether vermilion be pure or not. A small pinch should be heated over a spirit-lamp on a fragment of hard porcelain ; no appreciable residue will be found, unless red-lead, red iron oxide, brickdust, or other non-volatile adulterants be present. Carmine, which is sometimes added to scarlet vermilions to approximate their hue to that of the crimson varieties such as the Chinese, may be detected by laying a pinch of the powdered pigment on a small pad of white blotting-paper, and moistening the substance with a few drops of strong ammonia-water ; a crimson stain will appear on the paper if carmine or crimson-lake be present. The colour of a good vermilion is not changed by moistening it with nitric acid. The accidental impurities which impair the hue of vermilion are free sulphur, and compounds of iron and lead ; that prepared in the wet way often retains alkaline salts, owing to imperfect washing. A spurious vermilion, called anti-vermilion or antimony vermilion, is made by warming antimonious chloride with sodium thiosulphate.

Vermilion prepared from the mineral or native cinnabar is probably less liable to change than the artificial products, whether obtained by the dry way or the moist way ; but 'moist way' vermilions are certainly the most alterable. And it may also be remarked that the more finely a vermilion is ground, the less stable it is—at least, as a water-colour paint. Thus it happens that, other things being equal, an orange-vermilion is inferior in permanence to a scarlet, and a scarlet-vermilion to one inclining to crimson. As an oil-pigment, vermilion does not dry well, but suffers, especially if it be locked up in copal or paraffin, no change by light or impure air ; 100 parts of the dry substance require less than 20 parts of oil. In water-colour painting most vermilions are found to be changed on exposure, the solar rays gradually converting the red into the black modification of mercuric sulphide, without, of course, producing any chemical alteration. This change occurs even in the absence of air and of moisture. Impure air, *per se*, even if sulphuretted hydrogen be present, does not discolour vermilion.

Anyone who has examined old illuminated manuscripts must have noticed the apparent capriciousness with which the ornaments, and especially the initial letters, painted with vermilion, have been affected. I have more than once observed that, while all the vermilion used in one part of a missal or choral-book has remained red, a leaden hue has spread irregularly over the rest of the work in places where this pigment has been used. This may be due to a different sample of vermilion having been employed, but sometimes to a change in the technique, as a change in the style or handiwork is often associated with the difference above described. In oil-painting, there are no permanent pigments with which vermilion may not be safely mixed. Only when it contains impurities, such as free sulphur, does it darken flake-white.

Vermilion prepared from native cinnabar is found perfectly preserved in the flesh-tints of Italian tempera-paintings of the thirteenth and fourteenth and fifteenth centuries. It has stood in the wall-paintings of Pompeii, where it often seems to have been waxed. A comparatively recent but instructive instance of the permanence of vermilion in oil is furnished by a portrait, dated 1758, in the National Portrait Gallery. It represents the painter, Hogarth, with his palette set before him. The second of the dabs of colour thereon is vermilion, perfectly intact. In the same collection there is a portrait by Marc Gheeraedts of Mary Sidney, Countess of Pembroke, in which the vermilion has stood. This work was painted in 1614. Scores of earlier and later examples might be cited.

The variations in hue observable in different specimens of vermilion are mainly due to the differing degrees of fineness in which the pigment occurs. The coarsest grain corresponds with a crimson hue, and then we have every variety of colour ranging from scarlet to reddish orange or orange. The processes of regrinding and 'washing-over' enable us to obtain the kinds separately. And if we repeat these operations often enough, we may ultimately convert the whole of a crimson vermilion into the orange form. It was formerly supposed that the latter material was a mere scum, or impurity, or at least differed from the crimson kind in composition. When any vermilion is mixed in tint with white, an opposite effect to that of further grinding is produced. For, as the early writer Eraclius states: 'If you mix white with vermiculus, carmine is made'—that is, the hue of the mixture becomes more rosy, and therefore further removed from orange.

MADDER: *Pink Madder—Rose Madder—Madder Carmine—Madder Red—Rubens' Madder—Madder Purple—Madder Lake—Madder Brown—Carmin de Garance—Laque de Garance—Krapp-lack.*

Some authorities assert that madder was used in dyeing long before its employment in painting. But if the ancient Greeks and Romans were unacquainted with a red pigment derived from madder, yet there are good reasons for believing that such substances were known in Europe as early as the thirteenth century. Even in England, such a pigment is almost certainly referred to, under the name 'sinopis,' in the middle of the fourteenth century. Now Alcherius (close of fourteenth century) tells us that 'sinopis is a colour redder than vermilion, and it is made from varancia.' 'Varancia' is clearly garance—that is, madder—the same material being named 'warancia' and 'waranz' in a British Museum manuscript (Sloane, No. 416) which contains recipes of the fourteenth century. Besides 'sinopis' (strictly, a red earth), madder-lake was called, in English account-rolls of the fourteenth century, 'sinopre' and 'cynople.' It is, however, difficult, if not impossible, to ascertain the precise date at which pigments derived from madder came into use in the various schools of painting in Europe. For the nomenclature of pigments has always been somewhat vague, while the evidence furnished by existing pictures does not at present enable us to trace back the use of madder paints to an earlier time than the close of the fifteenth century. Eraclius does not mention madder, nor does Cennini, who lived at a much later time. Mr. R. Hendrie, in his notes to 'Theophilus,' speaks of an English manuscript of the fourteenth century in which directions are given for extracting the colouring matter of 'madyr.' From these directions we are, perhaps, justified in concluding that the preparation of a kind of liquid paint was intended.

The European madder-plant, a native of Greece, belongs to the tribe Galieæ, of the order Rubiaceæ; it is the *Rubia tinctorum* of Linnæus. Several other species of this genus are used or grown in India for the sake of the red dye they afford. Among such species, *Rubia cordifolia* (Linn.) and *R. sikkimensis* (Kurz.) may be named, but the European madder is also cultivated extensively in India. Much madder was formerly grown in the Levant, in Holland, and in the south of France; but the manufacture by artificial means from the anthracene of coal-tar of its two chief colouring matters, alizarin and purpurin, has almost entirely extinguished the cultivation of the madder-plant in Europe. We shall have something to say presently concerning the artificial products above named.

The root of madder contains a much larger proportion of the colouring matters (or, it would be more correct to say, colour-making substances) than the other parts of the plant. They occur dissolved in the yellow cell-contents of the soft tissue of the root. The finest madder was grown in the 'Palud,' a chalky valley near Vaucluse.

The colouring matters obtained from madder exist in the plant in the form of glucosides. These glucosides are resolved by the fermentation, brought about by a peculiar ferment in the plant itself, and by many chemical agents, such as mineral alkalies and acids, mainly into glucose on the one hand, and on the other, into the several colouring principles. Of such colouring principles the glucosides in madder yield at least three, of which the most important are these two :

1. Alizarin, $C_{14}H_8O_4$.
2. Purpurin, $C_{14}H_8O_5$.

Both alizarin and purpurin are now manufactured artificially from anthracene. This compound, which occurs in coal-tar, is a crystalline fluorescent hydrocarbon, $C_{14}H_{10}$. By a series of processes this substance gives rise to alizarin and purpurin, which are in all respects identical with these colouring matters as derived from the madder plant itself. The artificial alizarin of commerce contains several other colouring matters, two of which are better known than the others; these are anthrapurpurin ($C_{14}H_8O_5$) and purpuroxanthin ($C_{14}H_8O_4$). Purpuroxanthin is also present in the natural pigments derived from madder, but it exists in small proportion. Of all these compounds alizarin is the most important and the best known, and yields lakes having various hues of crimson, rose, purple, violet and marone, according to its purity, its concentration, and the nature of the base (alumina, iron-oxide, manganese oxide, copper oxide, or lime with alumina) with which it is associated. The purpurin, and anthrapurpurin resemble one another closely, and give pigments which are generally characterized by more orange or red hues than those obtained with alizarin. The rose and pink madders and the madder carmines of commerce are generally so manufactured as to include, for their colouring constituents, much alizarin and very little purpurin. A few indications of the ordinary methods of preparing these lakes may first be given.

The material used is often that called 'madder flowers,' which consists of the finely ground dried root after it has been submitted to the action of dilute sulphuric acid, and washed. Four pounds of this madder are taken and warmed for two or three hours on a steamer, with a solution of 1 pound of pure alum in 1 gallon of water. The mixture is placed in a filter-press, and the liquor obtained (which must be perfectly clear) precipitated by the gradual addition of a solution of sodium carbonate. The first portions of madder lake which fall, being the best, should be collected apart. All the precipitates should be thoroughly washed with rain or distilled water till the wash-waters are no longer troubled on the addition of barium chloride solution; they are then moulded into small cones, drops, or discs, and carefully dried at a moderate temperature. Another process for preparing madder lakes is a modification of the above. Four pounds of madder-root in powder, after having been fermented and then washed with a weak solution of sodium sulphate, are boiled for fifteen minutes with 4 gallons of a 10 per cent. solution of pure alum, the whole is filtered, and at a temperature of 45° partially neutralized with a solution in water of about 8 ounces of pure sodium carbonate. The liquor is now brought to the boiling point, the madder-lake which is then deposited is to be thoroughly washed and then dried: it is much denser than that produced by the preceding process. In order to prepare a madder carmine in which there is little earthy base, the following plan is frequently adopted: Some of the very finest madder root is fermented in the usual way, dried, ground, and treated with four times its weight of sulphuric acid, diluted till it shows a specific gravity of 1·6. The vessel in which the mixture is made is kept cold. After three hours some water is added to the mass, and then the whole is filtered through asbestos cloth. The filtered liquor is allowed to drop into a large quantity of rain or distilled water; the madder carmine will separate and collect at the bottom of the water. It is to be thrown on a filter, washed, and dried.

By the employment in various proportions of washed, moist aluminum hydrate, and by choosing various qualities of madder root, a number of hues and tints of rose and pink madder may be obtained when one or other of the methods above described is adopted. The oxides of iron, manganese and copper, when used in association with more or less alumina as a base for receiving the various colouring matters of madder, give other hues, including madder-purple and madder-brown.

But occasionally the pigments sold under these names are mixtures. For instance, burnt sienna and copper ferrocyanide have been found in samples of madder brown; the presence of copper in madder brown seems, however, to be usual, but it arises from the employment of copper sulphate in its preparation along with alum.

I select from a number of analyses of madder pigments the two following as illustrative of some peculiarities in the composition of these important preparations:

	Rose Madder	*Madder Carmine*
Water given off at 100° C.	12·0	13·0
Water, combined, etc.	28·1	15·0
Colouring matter	10·0	22·6
Alumina	49·9	49·4

According to some authorities the finest rose madders and madder carmines consist mainly of anthrapurpurin, associated with a base of alumina. But my own experiments with anthrapurpurin do not confirm this view. The following is a brief account of my chief results. The three colouring matters—anthrapurpurin, purpurin, and alizarin—were first prepared in a state of the greatest possible purity, the two last-named substances having been obtained from madder-root, the anthrapurpurin from anthracene. An alkaline solution of each of these three colouring matters was then made, ammonia being the solvent alkali in one set of trials, and caustic potash being used in another series. No more alkali was employed than was necessary to effect the solution of the substances. The various coloured liquids were then precipitated by means either of a solution of pure potash alum, or by boiling them with pure gelatinous aluminium hydrate, care being taken to use about the same proportion of alumina to colouring matter in each set of experiments. The lakes obtained were characterized by the following hues :

Anthrapurpurin lake ; red, nearly pure.
Purpurin lake ; red, tending towards orange.
Alizarin lake ; crimson red.

I must confess that even my alizarin lake was not of so pink or rosy a hue as that known in commerce as rose-madder and madder-carmine, which, when genuine, is certainly made directly from preparations of the madder root, and in which I have been unable to detect any colouring matters besides alizarin, although small quantities of fatty acids are occasionally present. The advantage of using for artistic purposes those madder lakes which have been prepared with alizarin is not confined to the beauty and purity of the colours they furnish, for experiment shows that these lakes are the least affected by light of all the pigments derived from madder. The manufacture of the above-named choice pigments for artists appears to be the only reason which prevents the cultivation of the *Rubia tinctorum* from being wholly extinguished through the energetic competition of the artificial chemical product.

From alizarin and from purpurin (either natural or artificial) madder lakes may be readily prepared by dissolving these substances in the smallest necessary quantity of an alkali, such as ammonia or sodium carbonate, and then adding a solution of a pure aluminium salt or some pure freshly precipitated and thoroughly washed aluminium hydrate. The best artificial alizarin of commerce occurs as a yellowish powder, presenting the aspect of raw sienna. It may, however, be obtained in yellow or orange red crystals, either by repeated crystallization from a solvent or by sublimation *in vacuo.* Its colour is always brighter than that of purpurin, which in powder has about the hue of Venetian red. But when solutions are made of these two substances in alkalies, then it is seen that the colours are reversed—alizarin yielding a crimson verging upon purple, and purpurin a red verging upon crimson. Differences of colour will be noticed in the lakes prepared with these two bodies. The directions for preparing pigments from the above-named bodies are practically identical with those already given in outline, but the minute details of manipulation can be learned only in actual practice. One peculiarity, however, distinguishing purpurin from alizarin, may be here noticed. It will be found that, while alizarin is practically insoluble in cold saturated alum-water, and but slightly soluble in the same liquid even when boiling, purpurin dissolves to a marked extent in this solvent. From this purpurin solution, which possesses a crimson colour and exhibits a fine orange red fluorescence, alkalies throw down a pure red-coloured lake. The following process gives a red madder of excellent hue: Equal weights of pure alum (absolutely free from iron and lime) and of the purest artificial purpurin in powder are ground together, and then washed with cold water until the washings are colourless; then the residue on the filter is boiled with a 5 per cent. solution of pure alum, filtered while boiling, and immediately neutralized with pure sodium carbonate solution (also boiling) until red flocks appear. These are filtered off, and constitute, when washed and dried, a fine pigment of a rich red hue. By heating the mother-liquor to 80°, and adding more sodium carbonate, a further and equally good product is obtained. The purpurin residue, when again heated with more alum-solution and precipitated as above directed, yields a further quantity. The final residue, after several such exhaustions, produces an impure madder lake, having a brownish red hue.

Although the madder colours are very much less affected by light than are the pigments derived from cochineal, yet it cannot be affirmed that any of them are absolutely permanent when continuously exposed. The following figures show approximately the amount and nature of the change, observed after certain intervals, in the case of several madder pigments used as water-colours:

Name of Pigment	*Original Intensity* = 10	*Change of Hue*
*Madder Carmine, A	After 1 year, 10	Very slight.
,, ,, B	,, 1 ,, 8	Much more purplish.
,, ,, C	,, 5 ,, 2	
,, ,, C	,, 7 ,, 0	
Madder Red	,, 1 ,, 6	Less red, more blue.
Rose Madder	,, 1 ,, 8	Slightly more purplish.
,, ,, B	,, 2 ,, 3	
,, ,, B	,, 5 ,, 1	Smoke-grey.
,, ,, B	,, 7 ,, 1	Grey.
Pink Madder	,, 2 ,, 1	
Purple Madder, A	,, 1 ,, 7	Duller, less red, more blue.
,, ,, C	,, 2 ,, 6	More bluish.
,, ,, D	,, 5 ,, 7	
,, ,, C	,, 7 ,, 2	
* ,, ,, E	,, 7 ,, 9	Somewhat puce
Brown Madder, A	,, 1 ,, 9	Less red, more yellow.
,, ,, B	,, 2 ,, 1	Grey.
,, ,, B	,, 5 ,, 1	
,, ,, B	,, 7 ,, 0	Grey.

The letters A to E indicate different samples of the several pigments, which were in all cases 'moist' colours; a parallel but less complete series with 'cake' colours gave practically the same results. The two samples marked * are instances of exceptional stability, and are of importance as showing the possibility of obtaining some, at all events, of the madder pigments in a satisfactory form. It is noticeable that the paler (pink and rose) madders, which contain much water, are generally more perishable than the concentrated madder carmine; the comparative trials having, of course, been made with washes of the same depth of tint.

MINERAL LAKE: *Pink-Colour—Laque Minérale.*

Attempts have been made to obtain mineral pigments of absolute permanence in order to acquire substitutes for the reds and purples of vegetable origin. None of them equals in intensity and splendour of colour the derivatives of madder. One of the best of these substitutes is mineral lake. This compound may be made by precipitating a solution of neutral potassium chromate by means of a solution of stannic chloride. The precipitate is collected on a filter and thoroughly washed. Still moist, it is ground into a paste with half its bulk of pure nitre, and allowed to dry. The dry mixture is projected, little by little, into some nitre heated to low redness in a crucible. When the basic chromate of tin has settled, the nitre, still fused, is poured off, and the residue washed thoroughly with water. The product thus obtained requires calcination for two hours at a high temperature in a luted crucible, in order to develop its colour which much resembles that of almond-blossom. It constitutes an unalterable pigment.

LIGHT RED: *Burnt Ochre—Rouge Anglais—Brun Rouge.*

Light red is, or ought to be, yellow ochre burnt—that is, calcined. The different varieties of yellow ochre yield, as might be expected, products having various hues and tints of this rather pale and dull brownish or orange red. Moreover, these hues depend in some measure upon the temperature at which the calcination is effected. To prepare light red, the selected yellow ochre is usually crushed and then roasted on an iron plate heated to redness. When the desired tint has been attained the material is thrown into cold water, ground, and washed. Light red may also be made by conducting the finely-divided yellow ochre suspended in a current of air into a heated chamber or furnace. Light red consists, then, of yellow ochre deprived of its water of hydration by means of heat. It is necessary to employ yellow ochre as free as possible from organic matter and from lime if a bright-coloured product be desired.

Light red possesses a considerable degree of opacity. Its hue may be defined as a scarlet, modified by a little yellow and grey. It is perfectly permanent and without action upon other pigments.

Light red boiled with hydrochloric acid will, if genuine, yield a solution, which after filtration will give no precipitate, but merely a slight cloudiness, on the addition of a few drops of barium chloride solution. The terms 'rouge Anglais' and 'Brun rouge' are not infrequently applied to artificially prepared iron reds.

VENETIAN RED: *Rouge—Crocus—Colcothar—Caput Mortuum Vitrioli.*

Originally Venetian red consisted of a native ferric oxide or red hæmatite, less purplish in its tints and washes than Indian red. But of recent years the name appears to have been transferred to a particular quality of artificial ferric oxide, made by calcining green vitriol. When this salt is heated in a crucible the upper portion of the product, which has been less strongly heated than the lower, is of a brighter red than the remainder, and after washing and grinding is sold as Venetian red. If moistened with a solution of nitre, again heated, and then ground and washed, the red tint of the product becomes somewhat brighter.

The hue of Venetian red is less brownish than that of light red, and not at all purplish like that of Indian red.

Venetian red, whether artificial or natural, is a permanent pigment which may be mixed with other permanent pigments without fear of injuring them; but it must be perfectly free from soluble salts and from any trace of sulphates. The presence of the latter may be detected by the test described under 'Light Red' and 'Indian Red.' But few commercial samples will stand this test, however, and we consequently find that many samples of Venetian red, owing to the presence therein of sulphates, exert an injurious action upon some of the organic pigments used as water-colours—notably, upon indigo.

A very fine native red ochre comes from the Banat, Hungary. It is represented by the formula $2Fe_2O_3$, H_2O, and goes under the mineralogical name of turgite. Its hue is that of a fine Venetian red, but it is probable that the fine native Indian reds and red ochres sometimes consist of or contain this hydrated ferric oxide, and are not really anhydrous.

INDIAN RED: *Persian Red—Indian Red Ochre.*

Indian red is a variety of red ochre, or red hæmatite, containing about 95 per cent. of ferric oxide, and having a slightly purplish hue. It varies somewhat in quality, and often requires sifting through a fine silk sieve, followed by washing over, in order to fit it for use as a pigment. Most of the Indian red imported from India is a natural product, but some has been prepared by calcination. Some so-called Indian red is imported from Ormuz in the Persian Gulf; some is an English hæmatite from the Forest of Dean.

Indian red, when genuine, is a perfectly permanent pigment in all media, and is without action upon other colours. It was extensively employed by the older masters of the English Water-Colour School, in association with true ultramarine, with Prussian blue, with indigo, or with indigo and yellow ochre, to produce the lilac greys of stormy clouds. The indigo in some of these greys having often perished, the Indian red (and the yellow ochre where employed) remains intact, giving a hot and frequently foxy red to spaces which were originally cool in hue, and comparatively neutral. This change has been incorrectly attributed to an action exerted upon the indigo by the

Indian red. But as indigo disappears when used alone, or when a thin wash of it on a sheet of gelatine is placed over, but not in contact with, a wash of Indian red, the current explanation of the phenomenon in question cannot be true. Greys made with light red or Venetian red show similar alterations of colour. Colcothar, or jewellers' rouge, the red oxide of iron obtained as a residue when green vitriol (ferrous sulphate) is calcined, has sometimes been called Indian red, and substituted for the native oxide. Those portions of the above-named residue which have been more strongly heated generally present something of the purplish red hue which belongs to the true native Indian red. And this peculiar hue may be imparted to ordinary rouge by moistening it with a weak solution of potassium chlorate, drying, and then calcining the mass once more. It generally contains basic ferric sulphate, and then should be looked upon with suspicion, for it may seriously damage the indigo and other organic pigments with which it is associated. If a small pinch of Indian red be boiled with hydrochloric acid, thrown on a filter, and the filtrate tested with barium chloride solution, the genuineness of the pigment will be proved by the absence of any white precipitate of barium sulphate.

RED OCHRE: *Red Hæmatite—Red Iron Ore—Scarlet Ochre—Red Chalk—Ruddle—Bole—Sinoper—Sinopis—Rubrica—Miltos—Terra Rosa.*

The pigments above-named are native ferric oxide (or iron peroxide) associated with variable proportions of mineral impurities such as clay, chalk, and silica. They differ from the yellow and brown ochres described on page 126, by not containing combined water, in other words, the iron to which they owe their colour is ferric oxide, not ferric hydrate. They occur in very many localities accompanying or even constituting some of the most important iron ores. Their colour varies with their physical state, and with their purity; some are iron grey, or even black, until they are finely ground, when they assume a cherry-red hue. Cappadocia yielded to the ancient Greeks some at least of their *sinopis*, or red ochre, but they were familiar with the process of calcining yellow ochre in order to redden its hue; and they thus prepared the pigment to which the name 'light red' is now assigned.

An unusually bright red variety of hæmatite from Cumberland gave me on analysis, in 100 parts, ferric oxide, 94·7; alumina, 2·0; silica, 2·2, and moisture, 1·1. But some almost equally rich red ochres contain much less iron oxide, a 'sinopis' from Anatolia, analysed by Klaproth, having been found to contain 21 per cent. only, and others, from other localities, not above 40. The paler varieties of a reddish brown or yellow brown hue often consist of mixtures of hæmatite and yellow ochre. A very fine red ochre from Tuscany, the Banat, Hungary, and from the Urals, contains about 5 per cent. of water, and is considered to belong to a distinct mineral species called *turgite*.

When red chalk (from Hunstanton, Norfolk) is calcined at a high temperature it loses its red colour, and becomes of a dull olive green hue, a change due in this case to the production of calcium ferrite, a compound of lime and ferric oxide.

The terra rosa of Italy owes its pinkish red hue to ferric oxide, but it is probably often, if not always, an artificial product.

All the pigments described under the title 'red ochre' are permanent, and without action on other pigments.

For the substitution of artificially prepared ferric oxide, or colcothar, for red ochre, and for the method of detecting it, reference should be made to page 167, under 'Indian red.'

INDIAN LAKE: *Lac Lake—Lack-Lack.*

Lac is a resinous secretion produced by certain plants when punctured by the larvæ of the Coccus lacca, an East Indian hemipterous insect. Amongst the trees which the insect chiefly attacks are *Butea frondosa*, *Ficus religiosa*, and *F. bengalensis*, *Schleichera trijuga*, *Shorea robusta*, and *Zizyphus jujuba*. The lac, though a secretion primarily derived from the tree on which the insects feed, is yet profoundly modified, particularly with respect to its colour, during its passage through the animal's body. It varies in colour with the species of tree, but always consists mainly of three substances, namely, a resin, a colouring matter, and a wax. The resin, which constitutes two-thirds of the substance, is obtained in the form known as 'seed-lac' by pounding in water the lac which has been removed by pressing with a roller the encrusted twigs on a floor. The water becomes red; from it, by evaporation, the crude 'lac-dye' is obtained. This is made into cakes, and dried.

The lac-dye of commerce contains nitrogenous and mineral matters, as well as several dark-coloured impurities, and some resin. In order to prepare a lake from it it should be first powdered, and then digested in spirits of turpentine, or benzene. The purified residue is, when dry, extracted with sodium carbonate solution: the liquor is afterwards filtered and precipitated with alum solution. The precipitate thus formed is washed and dried in the dark.

Indian lake was used by the Venetian and Flemish painters of the sixteenth century, but it does not seem, so far as one can gather from the notices of it in the work of De Mayerne, and in the Secreet-Boeck, to have been often obtained of good colour. It is even spoken of as a 'light brown.'

Indian lake is inferior in beauty, and in variety of hues, to the colours from madder; it is also more affected by light. But it is much less fugitive than crimson lake and the other cochineal pigments.

I do not know from direct experiment whether the comparative stability of the red colouring matter from *kermes* when used as a dye for animal fibres belongs also to all the lakes prepared from this substance: anyhow kermes pigments are not met with in commerce at the present time.

CARMINE AND THE COCHINEAL LAKES: *Carmin—Laque Cramoisie, or Crimson Lake—Purple Lake, etc.*

Cochineal consists of the dried wingless females of a species of coccus (C. cacti) which feeds upon several kinds of *Opuntia*, or cactus. The best quality comes from Teneriffe, and contains about half its weight of colouring-matter. This colouring-matter is a glucoside, to which the name of carminic acid has been given. When this substance is treated with weak sulphuric acid, it is resolved into a sugar and another colouring-matter called carmine-red.

The carmine of commerce is prepared directly from cochineal, and is the most concentrated and purest form of any of the pigments derived from this source. The methods of preparation differ, but in all the colouring-matter is extracted from the insects by means of boiling water, with the subsequent addition of small quantities of alum, or nitre, or potassium oxalate, or cream of tartar; occasionally a small quantity of stannous chloride is employed also. The liquor, after a repose of some days or even weeks, deposits a great part of the colouring-matter as a deep crimson-red powder, which is then thrown on a filter, washed and dried in the dark. However prepared, it contains fat, albuminoid matter, mineral salts, and other impurities; the finest varieties, however, dissolve perfectly in strong liquor ammoniæ. The liquor, which has deposited the carmine, gives up the remainder of its colouring-matter to freshly-precipitated aluminium hydrates, or, after having been rendered alkaline by potassium carbonate, to a solution of alum. Lakes are thus formed. Lakes are also made by directly precipitating cochineal extract with solutions of potassium carbonate and alum. A purplish tinge is given to the product by a small quantity of lime; other hues, generally dull, are imparted by the presence of iron, manganese, or copper in the solutions employed. The cochineal lakes always contain a larger quantity of water and of alumina (or other mineral basis) than carmine, and are consequently weaker. Scarlet lake is a mere mixture of crimson lake and vermilion.

Beautiful and rich as are the colours prepared from cochineal, not one of them should ever find a place upon the palette of the artist. They all become brownish, and ultimately almost disappear after a short exposure to sunlight or the more prolonged attack of strong diffused daylight. In six hours of sunshine a strong wash of fine crimson lake on Whatman paper lost 8 per cent. of its original intensity; this was on April 12. The loss during a second period of six hours' exposure was much less, but after the lapse of four months less than 5 per cent. of the original colour remained. In the case of carmine, from one to two years was required for the complete obliteration of every trace of the original crimson from a deep wash of this pigment. All the cochineal pigments become somewhat brownish during the course of fading, but ultimately, when all the red has disappeared, either a greenish grey or a faint sepia-like brown is the sole residue.

The term 'lake' belongs to all colouring-matters thrown down upon such a basis as alumina; but when purple, crimson, or scarlet is prefixed to the word lake, cochineal colours are always understood. So carmine used alone refers to cochineal carmine, although it is a general term for a group of rich pigments, of which madder carmine and indigo carmine are perfectly distinct examples, derived from madder and indigo respectively.

It is not necessary to say more about the various cochineal pigments, nor to point out their many falsifications, for their value as artists' colours is very small. No artist who cares for his work and hopes for its permanency should employ them.

Burnt Carmine.

This preparation should rather be called 'roasted carmine.' It is obtained by carefully heating the carmine made from cochineal. It possesses a beautiful hue, but is quite as fugitive as the product which yields it. Two years' exposure to sunlight completely destroyed a strong wash of cake burnt carmine on paper. The moist pigment had lost nine-tenths of its intensity at the end of the same period, while the small residual proportion of the colour had suffered no further change in depth at the end of a further lapse of three years. Burnt carmine is rather less fugitive in oil than in water colour. Experiments in the latter medium gave, after exposure to sunlight, the following residual intensities out of ten:

Moist, after two years	1
Cake, after two years	0
Moist, after five years	1

A sample of burnt carmine purchased of Messrs. Newman about the year 1815 was found (as might have been anticipated) to have retained its colour in the cake perfectly to the present year; but a wash of it on paper possessed no greater nor less degree of permanency than a wash of the same pigment prepared by the same makers in 1886.

RED LEAD: *Minium—Saturnine—Mine Rouge, Mine Orange—Mennige—Rosso Saturno.*

This beautiful orange-red pigment approaches in composition a compound of two molecules of protoxide of lead with one molecule of binoxide, and may be approximately represented by the formula Pb_3O_4. The paler and more orange-tinted varieties contain an excess of protoxide of lead, often accompanied by a little carbonate.

This pigment is peculiarly liable to discolouration in the presence of sulphuretted hydrogen: it acts energetically upon some paints, on the cadmium yellows, for example. It is quite inadmissible as a water-colour, and cannot be considered as safe in oil.

COBALT VIOLET: *Rose de Cobalt.*

This pigment consists of the oxides of magnesium and cobalt. It is prepared at a high temperature, and is quite permanent. Its hue is sometimes purple rather than violet, and it may then be used as a substitute for purple madder; but the colour of different preparations varies to some extent.

CHAPTER XVI

GREEN PIGMENTS

TERRE VERTE: *Green Earth—Terre de Verone—Grüne Erde—Terra Verde.*

THERE are two rather indefinite minerals, probably not really distinct, namely, glauconite and celadonite, which furnish the raw material from which the artists' pigment, generally known as terre verte, is prepared. The form, or variety, celadonite is the rarer of the two; it is softer than glauconite: both minerals are probably mixtures. Green earth of very good quality is found at Bentonico, to the north of Monte Baldo, near Verona, where it occurs in cavities in an amygdaloid rock. The best samples possess a deep olive-green colour; inferior specimens are celandine or apple-green. Green earth is obtained from a large number of European and American localities, and varies much in chemical composition. From its greenish hue it has been assumed to consist chiefly of a ferrous silicate—that is, a silicate of protoxide of iron, and in this way it is represented in the older analyses. But more exact analyses have shown that green earth contains but a small part of its iron in the protoxide condition, and that it is mainly a ferric silicate. A choice specimen of fine hue from Monte Baldo gave the following results on analysis, the numbers representing parts in one hundred:

Water, given off at 100° C.	4·1
Water, given off at a red heat	4·2
Ferric oxide (Fe_2O_3)	20·3
Ferrous oxide (FeO)	2·6
Alumina	1·7
Lime	1·1
Magnesia	5·6
Potash	6·4
Soda	2·3
Silica	52·0

Green earth is thus allied to the hornblendes, differing mainly in the partial replacement of soda by potash and in the presence of water. Being itself an alteration-product, it is not likely to be amenable to further change, particularly as the iron in it is for the most part fully oxidised.

Terre verte occurs in ancient Roman wall-paintings (the prepared pigment was found in the ruins of Pompeii), and was largely employed by the early artists of Italy in their works in tempera, fresco, and oil. But amongst the green pigments found in ancient wall-paintings in Rome and Pompeii there occurs also a substance of a richer and deeper hue than that possessed by terre verte. It was made by grinding into fine powder a kind of green jasper; it has proved quite permanent.

Terre verte is prepared by carefully selecting the richest-coloured and most uniform specimens of the mineral, grinding them to fine powder, and washing the pulverized material with rain-water; it is then dried. Sometimes the selected fragments are heated, and then quenched in very dilute hydrochloric acid to remove ochre and other impurities; the undissolved portion is then ground, washed thoroughly, and dried. Most samples of terre verte are found to be perfectly stable both in water-colour and oil-painting. It is a semi-opaque or translucent pigment, without much body in oil. It has no action on, nor is it affected by, other permanent pigments. When used in oil or tempera as a ground-colour or in the under-painting, terre verte sometimes becomes more conspicuous in the course of time, owing partly to the deepening of its own hue and partly to the increased translucency of the pigments which have been laid over it. Some samples of terre verte seem to be liable to become slightly rusty when brought into contact with lime hydrate in true fresco painting. This is probably due to the further oxidation of some of the ferrous oxide they contain.

Terre verte is rarely adulterated. A pure sample is not affected when drenched with liquor ammoniæ, becoming neither more bluish (presence of copper compounds), nor more brownish (presence of Prussian blue).

GREEN OXIDE OF CHROMIUM: *Chromium Sesquioxide—True Chrome Green—Opaque Oxide of Chromium—Vert de Chrome—Grünes Chromoxyd.*

The so called native oxide of chromium, or chrome ochre, is a mere greenish clay, containing not more than 10 per cent. of chromium oxide. But another mineral (from Okhansk, in Siberia) is of a deeper green, and sometimes contains one-third of its weight of this oxide. We are not aware that any attempt to employ these minerals as pigments has ever been made.

The green oxide of chromium, Cr_2O_3, varies in hue, in depth of colour, and in opacity according to the process of its preparation. It may be made either in the 'dry' way or the 'wet' way. Perhaps the finest quality is that produced by heating mercurous chromate in a retort till the whole of the mercury has been distilled off. This is an expensive process, but the product is excellent in colour and body.

Ammonium bichromate gently heated, in small portions at a time, yields a dull-coloured but a useful variety of this pigment; but there is a similar, though more economical process, yielding a better-coloured product. This consists in gently heating together and then calcining a mixture of 3 parts of neutral potassium chromate with 2 parts of ammonium chloride. The mass is to be thoroughly washed with hot water, dried, and again ignited. Two other processes for preparing this pigment in the dry way may be commended. In both of them potassium bichromate (free from iron) is used, but in the one case sulphur, in the other starch, is employed, in order to effect the reduction of the chromic salt to the condition of sesquioxide. The sulphur method yields at once a good pigment, which needs nothing but a thorough washing, first with very dilute sulphuric acid and then with water, followed by grinding, to fit it for use, but a second calcination is required in the starch process. In order to prepare oxide of chromium by the wet way, a solution of an alkaline chromate or bichromate is to be reduced by sulphur or other reducing agents, or a sesquisalt of chromium is precipitated by ammonia or a fixed alkali. The hydrated sesquioxide is thrown down; after having been washed, this substance is gently calcined. The hydrate itself, when air-dried, constitutes one of the forms of 'transparent oxide of chromium.'

The ordinary or opaque oxide of chromium usually occurs in the form of a greyish-green powder of considerable body. It is quite permanent under all conditions of exposure and of commixture with other pigments, and is available in all the processes of painting. Its tints with flake-white, and the yellowish greens which it yields with aureolin, are peculiarly valuable to the landscape-painter.

Green oxide of chromium is rather imitated than adulterated. This pigment when genuine is not altered in colour either by ammonium sulphide or caustic potash. But under the name of chrome-green mixtures of Prussian blue and lead chromate are constantly sold. These preparations generally contain some barium sulphate, often a considerable quantity of gypsum, and, occasionally, alumina.

VIRIDIAN: *Emerald Oxide of Chromium—Vert Pannetier—Vert de Guignet—Vert Emeraude—Mittler's Green.*

About the year 1838 Pannetier and Binet began to make a beautiful chromium green by means of a secret method. Many years afterwards M. Guignet discovered and patented a process by which this admirable pigment could be manufactured. About 8 parts of crystallized boracic acid and 3 parts of potassium bichromate are thoroughly mixed and calcined. The mass obtained is treated with cold water, washed by decantation, ground wet, washed with hot water, and carefully dried. The product is a hydrated chromium sesquioxide, in which a variable amount of the boracic constituent is often retained. Viridian is, however, essentially a hydrated sesquioxide of chromium, having the formula Cr_2O_3, $2H_2O$.

It would be difficult to exaggerate the value of this addition to the artist's palette. The colour of viridian is a very deep bluish green of great purity and transparency. It furnishes, with aureolin on the one hand and with ultramarine on the other, an immense number of beautiful hues, adapted to represent the colours of vegetation and of water. It is quite unaffected by sunlight and sulphuretted hydrogen, and it has no action on other pigments. Moreover, it may be safely used with all the different painting media, and upon all kinds of painting-grounds.

Viridian is perhaps the name by which this pigment is

now best known in England. It is unfortunate that it should be called 'Vert Émeraude' in France, since it has little in common with the poisonous emerald green of our colourmen. This pigment is sometimes adulterated with baryta yellow in order to modify its hue. This substance may be readily detected by moistening the pigment with dilute hydrochloric acid, when a yellow solution is obtained.

Arnaudon's chrome green is of a somewhat opaque green hue, rather like that of Schweinfurt; it is a phosphate of chromium. A similar product is Mathieu-Plessy's chrome green; but under the name of 'chrome green' ('Grüner Zinnober') spurious pigments are constantly sold, the commonest of them consisting of mixtures of chrome yellow and Prussian blue. They are worthless in comparison with viridian.

COBALT GREEN: *Rinmann's Green—Vert de Cobalt—Vert de Zinc—Kobalt Grün.*

It has long been known that the oxide or a salt of zinc, moistened with a solution of cobalt nitrate, and then strongly heated before the blow pipe, gives a porous mass of a beautiful green hue. This compound or mixture of the oxides of zinc and cobalt may be prepared by: (1) Precipitating with an alkaline carbonate a mixture of the nitrates of cobalt and of zinc, and then strongly heating (after washing) the precipitate formed; (2) Making a paste of zinc oxide and water, and adding to it a solution of nitrate or sulphate of cobalt, or of roseo-cobaltic chloride; the mass is then dried, calcined at a dull red heat, thrown into water, ground, washed, and dried. Method No. 2 gives a finely-coloured product, the depth of hue being proportional to the percentage of cobalt oxide. If the latter oxide amount to one-third of that of zinc the colour is a very deep bluish green; with no more than one-sixth the colour is still rich. Some specimens do not contain more than one-twentieth—sometimes even less—of cobalt oxide, and yet they are far from pale. An excellent deep sample contained 12 per cent. of cobalt oxide.

When properly prepared, cobalt green is a pigment of great beauty and power.* The deeper tones of cobalt green are almost transparent in oil. The pigment works well, is quite permanent, and has no action on other pigments. Cobalt green is, in fact, one of the too-rare pigments which is at once chemically and artistically perfect. It must be admitted, however, that it may be exactly imitated by a mixture of viridian and artificial ultramarine with a little zinc white.

Cobalt green is again coming into artistic use, as it is equally adapted for all the methods of painting. It was discredited for awhile by the inferiority of the product obtained by Rinmann's original process (No. 1 above). It ought not to be an expensive pigment.

Sometimes cobalt green is prepared by precipitating a cobalt salt with an alkaline arseniate or phosphate, and then heating the precipitate with zinc white.

EMERALD GREEN: *Cupric Aceto-Arsenite—Schweinfurt Green —Vert Paul Véronèse.*

This pigment was discovered in 1814 during the course of experiments made with the object of preparing an improved Scheele's green. It may be prepared by half a dozen slightly differing processes, but in all verdigris (or vinegar and blue vitriol) and white arsenic are the two essential materials employed. Generally verdigris is dissolved in acetic acid, and added to a boiling aqueous solution of white arsenic:

* The cobalt green sold by Edouard, of Paris, is by far the best I have examined.

on continued ebullition a deposit of emerald green occurs. Sometimes copper sulphate, potassium arsenite, and acetic acid are employed. Whatever the method, it is necessary that the coloured product be washed with boiling water to remove the last traces of soluble salts.

The hue of this pigment is a nearly normal green, slightly verging upon bluish green ; it is brighter and more opaque than Scheele's green, and, like it, is a deadly poison. It is less attacked by sulphuretted hydrogen than Scheele's green, but, as a water-colour, does not long remain untarnished in impure air. In oil it is practically permanent, both alone and when used with the majority of permanent pigments. It is, however, quickly blackened by the cadmium yellows. Emerald green cannot be relied upon as permanent in fresco and tempera painting. Its use in wall-papers and in the decoration of all domestic furniture and fabrics is to be deprecated, by reason of its poisonous character; but is, happily, at the present time, in great measure abandoned.

Emerald green, if pure, dissolves perfectly in boiling dilute nitric or hydrochloric acid ; the solutions thus made should yield no precipitate with a few drops of barium chloride solution. An undissolved residue generally shows the presence of baryta white.

Emerald green may be distinguished from Scheele's green by a simple experiment. If a small pinch of the dry powder be warmed with a few drops of moderately-strong sulphuric acid (half oil of vitriol, half water), acid vapours, having the smell of vinegar, will be given off only when true emerald green is the subject of the experiment.

On the Continent, Vert Émeraude is the name given to viridian, the emerald oxide of chromium.

SCHEELE'S GREEN: *Cupric Arsenite — Swedish Green — Mittis Green—Scheele's Grün.*

This pigment, discovered in 1778, is an arsenite of copper with an excess of copper oxide. It is best prepared by dissolving, in separate portions of hot water, white arsenic and blue vitriol. The solutions are then mixed, and to the mixture is added, in small successive portions, a solution of potassium carbonate. These additions are stopped when the precipitated pigment has attained its maximum of colour intensity. In another process, a hot solution of potassium arsenite is added to a hot solution of blue vitriol. This pigment needs thorough washing with hot water, and must be dried at a moderate temperature.

Scheele's green presents nearly the same characteristics as emerald green, but is in every way inferior to that pigment. It is eminently poisonous. It should not find a place on the palette of the artist.

Vienna green, Mittis green, and Veronese green are names which have been given to specially prepared varieties of this cupric arsenite; but there are very many other designations by which pigments of essentially the same composition are known. They are prepared by slightly modified processes, and frequently contain such foreign matters as chalk, heavy spar, or gypsum.

MALACHITE: *Green Verditer — Green Bice — Mountain Green—Green Carbonate of Copper—Vert de Montagne —Berggrün.*

This green copper mineral was employed as a paint by the ancients. It often accompanies the blue carbonate, and occurs in many European, Asiatic, African, and American localities. The mines at Ekaterinburg and

Nischne Tagilsk in Russia, and at Burra-Burra in South Australia, furnish malachite of fine colour; it also occurs abundantly in Namaqualand. Its variations in depth of colour are due less to impurities than to differences in its state of aggregation. Its specific gravity is about 4. Its composition is that of a hydrate and carbonate of copper. It may be represented by the formula $CuCO_3,CuH_2O_2$. It contains, therefore, less of the copper carbonate than azurite, or blue verditer.

Malachite requires no other treatment than careful grinding to fit it for use as an artists' pigment. The raw material must, however, be carefully selected, and all visible impurities, such as ochreous veins and deposits, and azurite, completely removed. An artificial malachite has been prepared, and is often substituted for the mineral; but it is inferior in colour and stability to the native form.

Malachite as an oil-paint has often proved to be permanent, although it may seem to acquire a dull, brownish hue, owing to the darkening and yellowing of the oil; sometimes, however, it becomes somewhat olive in colour. In admixture with cadmium yellow it is liable to blacken. It is so easily injured by impure air when unprotected by any hydrofuge, that it is quite inadmissible as a water-colour. In old tempera paintings it is sometimes found to have stood well; but the sulphur from the egg-medium and from the size has not infrequently browned it.

Malachite is sometimes adulterated with baryta white; sometimes a mixture of that pigment and an arsenical green is substituted for it. The former falsification may be detected by boiling the sample in hydrochloric acid, when the malachite dissolves, leaving the baryta white as a sediment. To detect an arsenical green, a small portion of the sample should be mixed with powdered charcoal, and heated in a long narrow test-tube; a dark sublimate of metallic arsenic will form on the cooler part of the tube.

VERDIGRIS: *Basic Copper Acetate—Vert-de-Gris—Verdet de Montpellier—Grünspan.*

This green copper pigment was called by the writers of the fourteenth century 'viride Græcum,' or, more simply, 'viride,' 'viride terrestre' being used for green-earth—that is, 'terre verte.' 'Vert-de-Grèce'—that is, verdigris—was used by the ancient Romans as a pigment, and has been detected in the wall-paintings of Pompeii. It occurs in early Italian tempera pictures; but it has frequently injured the gesso-ground on which it has been laid, forming calcium acetate with the calcium carbonate, and disintegrating the surface. The blackness of the shaded parts in many Venetian and Spanish pictures of the sixteenth and seventeenth centuries has been attributed to the changes which this pigment suffers in oil. The mediæval writers on the practice of painting endeavoured to show how the peculiar liability of verdigris to change could be obviated by locking it up in some hydrofuge substance, such as a resin or balsam. Later authorities have recommended copaiba for this purpose; but copal varnish and hard paraffin are more effective. The large proportion of resinous matter employed by early painters for the protection of verdigris from alteration, and the success of this precautionary measure, may be seen in the green drapery in several pictures by Van Eyck and Mabuse in the National Gallery and at Hampton Court. It should, however, be stated that the older processes for preparing verdigris often yielded a product much more alterable in the presence of damp than is the verdigris which for a century or more has been made at Montpellier.

Verdigris is commonly called in chemical language a

basic acetate of copper. In fact, it is a mixture of three such acetates, its varying hues, ranging from green to greenish blue, being dependent upon the relative proportions of these acetates. The most blue basic acetate contains 1 molecule of copper acetate, and 1 of copper hydrate, with 5 molecules of water; the greenest has twice as much acetate. Average verdigris contains in 100 parts about 29 parts of anhydrous acetic acid, 43 of copper oxide, and 27 of water. It is nearly insoluble in cold water; but by continuous washing, or by continuous exposure to moist air, is ultimately decomposed.

The Montpellier process for making verdigris consists in exposing thin strips of metallic copper to the vapours arising from grape marc undergoing the acetic fermentation. The operation is conducted in a moist, warm atmosphere; finally, the whole substance of the metallic copper is transformed into verdigris.

An impure atmosphere containing sulphuretted hydrogen blackens verdigris; it is also affected by moisture and by carbonic acid. As a water-colour, it is quite inadmissible; in oil, it stands pretty well if 'locked up' in the way already described. But it acts energetically upon several important pigments, and is very poisonous. For these reasons its employment in artistic painting ought to be abandoned.

Verdigris, if pure, dissolves perfectly in liquor ammoniæ, any gypsum or barytes present as diluents or adulterants remaining undissolved. If blue vitriol has been added to verdigris, it also will dissolve in the ammonia; but this falsification may be detected by acidifying the ammoniacal solution with hydrochloric acid, and then adding solution of barium chloride—a white precipitate of barium sulphate indicates the presence of copper sulphate.

CHAPTER XVII

BLUE PIGMENTS

ULTRAMARINE: *Lapis Lazuli Blau—Lasurstein Blau—Outremer—Bleu d'Azur.*

THERE are at least three mineral species, closely allied in chemical composition, and generally presenting a more or less marked blue colour, which contain as their essential constituents the five elements, silicon, aluminium, sodium, sulphur, and oxygen, and which owe their characteristic hue to the same compound. From one of these minerals, a variety of 'hauyne,' often called 'lapis-lazuli,' the true or native ultramarine is obtained. This stone occurs, of very varying purity and colour, at Bucherei, Transbaikal, and in many other Siberian localities; at Ditro, in Transylvania; in the Andes of Ovalla, Rio Grande; and in several regions of Persia, Tibet, and China. It is the 'sapphire' of ancient authors. Small golden specks of iron-pyrites are frequently irregularly scattered through its substance; it is also very frequently associated intimately with portions of the rocky gangue, or matrix (limestone, syenite, granite, etc.), in which it occurs. Very fine lapis-lazuli comes from Tibet.

To prepare a pigment from this mineral, selected pieces of small size, as free as possible from pyrites or other impurities, are heated in a crucible and quenched (étonné) in cold water, or very weak vinegar. The material, thus disintegrated, is washed by decantation, and then dried and carefully ground. The powaer is then purified by elutriation, or 'washing over,' the several wash-waters depositing pigments of different depths of colour, and of different degrees of fineness. Some manufacturers adopt an old process, and make the powder into a soft mass with a little rosin, linseed oil and bees'-wax, and knead, beat, or macerate the lump under very weak potash, or soda-lye—the alkaline water carries off or withdraws the greater part of the pigment, and deposits it on standing. The richness of the blue product obtained depends primarily upon the original quality of the stone, but several grades are always procurable from the same raw material by means of the above-described processes, bluish grey and grey powders, known respectively as ultramarine ash and mineral grey, being the last and the least valuable products, while the deepest and finest pigments are deposited from the earliest wash-waters.

Optically, the superb blue of native ultramarine approaches more closely than the blue of any other pigment to the pure normal blue of the solar spectrum ; it shows very little violet, and in this respect is unlike most specimens of artificial ultramarine. Ultramarine is somewhat harsh and granular in texture, a characteristic which may be reduced by a small admixture of Chinese white, but which becomes more marked when it is used, as a light wash, or in conjunction with transparent pigments, in water-colour painting. It more nearly approaches transparency when used in oil, and is then of excellent working quality.

Ultramarine withstands the action of light, moisture and sulphuretted hydrogen perfectly : it neither affects nor is affected by any other pigments. It dries well in oil. It is decolourized at once by a hot solution of alum, and *ultimately* even by a saturated cold solution, which, however, bleaches very quickly all but the most stable varieties of the artificial pigment. All mineral acids, save carbonic, and all the common organic acids, such as acetic, oxalic and citric, discharge the colour of native ultramarine. It is only by a combination of several tests in the hands of a skilled chemist that the discrimination of genuine from spurious ultramarine may be with certainty accomplished.

In the works—both in fresco and tempera, and in oil—of many of the old masters, and in a large number of illuminated manuscripts, the permanence of true ultramarine may be seen. If in some cases it has acquired a greenish or dull hue in oil-paintings, such change is due to the yellowing of the oil and varnish, and not to any deterioration of the pigment.

The price of genuine ultramarine is very high. This is due less to the scarcity of the original lapis-lazuli from which it is derived, than to the small yield and to the elaborate and tedious operations by means of which the pigment is prepared. But when every allowance is made on account of the troublesome and lengthy process of manufacture, the cost of ultramarine is unwarrantably excessive.

ARTIFICIAL ULTRAMARINE: *New Blue—French Blue—Permanent Blue—Gmelin's Blue—Guimet's Blue.*

In the year 1814 a blue colouration, subsequently proved to be due to ultramarine, was noticed in the soda (black-ash) furnaces of St. Gobain. About fourteen years afterwards a method of making the same blue substance at will was discovered by Christian Gmelin, and by Guimet; this method was founded in part upon chemical analyses of natural ultramarine, and in part upon a study of the conditions under which the above-named blue colouration occurred. By successive improvements in its manufacture artificial ultramarine is now produced at a cost of no more than a few pence per pound. It is chiefly made in Germany and in France.

The raw materials employed in the preparation of ultramarine are, kaolin, or China-clay, silica, sodium sulphate, sodium carbonate, sulphur, charcoal and rosin. Some makers omit the sodium sulphate, others the rosin, while calcined alum is occasionally substituted for the kaolin. These materials are heated together in closed crucibles in a furnace, and slowly cooled. A greenish porous cake is the product: this is powdered and gently roasted, after the addition of a little sulphur, for some hours. The material is again powdered, and then washed and dried: further calcination is sometimes required to develope the proper blue colour.

In preparing artificial ultramarine for use as an artists' pigment it must be very finely ground, and very thoroughly washed with water free from lime. The grinding not only improves the colour, but renders working with the paint less difficult to manage; the washing removes soluble sulphates and certain sulphur-compounds, which are liable to discolour some of the pigments (those containing lead or copper) with which the ultramarine may be associated in a picture.

The hue of artificial ultramarine is commonly of a less pure quality than that of the natural pigment, verging somewhat towards a purple. But its range of hue is considerable, from a greenish-blue to a decided violet. The greenish-blue and blue varieties are not affected in hue by admixture with zinc white, but the varieties which incline towards violet become remarkably enfeebled in richness of colour by this admixture, such weakening being out of all proportion with the dilution of tint which would be expected to ensue from this addition of white. However, other white substances do not produce this curious result.

Weak acetic acid, and a saturated cold solution of alum, which are without immediate action upon natural ultramarine, generally change the hue, and always ultimately decolourize the artificial product. Those kinds which have a somewhat violet tinge resist the destructive action of the above re-agents longer than the pure blue and greenish-blue varieties. Neither sulphuretted hydrogen, nor caustic lime or other alkaline substance, affects the colour of artificial ultramarine.

Although the colour of ultramarine is certainly due to a substance containing sulphur, the precise chemical composition of this blue substance has not yet been determined. Some chemists are of opinion that it is a compound of aluminium, sodium, sulphur, and oxygen; others regard it as a sulphide of aluminium. There is little doubt, however, that it contains sulphur in two conditions. Some curious derivatives of blue ultramarine, of various colours, have been obtained, in which it is believed that the sodium of the original compound has been replaced by other metals; such is the yellow 'silver-ultramarine,' prepared by keeping blue ultramarine in a solution of silver nitrate. These bodies are, however, useless as pigments. The function of the silica in ultramarine is not known, although it forms from 30 to 45 per cent. of the total weight of all the varieties, and although it has been found by experience that ultramarines rich in silica resist the action of alum better than those which are poor in this constituent: these silicious ultramarines are sometimes sold under the name of 'Oriental Blue.'

Artificial ultramarine, when properly prepared, is permanent both in water and oils. When thin washes on paper appear to lose strength as they dry, or soon afterwards, the change is due to the chemical action of the alum, or other aluminium compound, present in the size of the paper. If an ultramarine should discolour emerald green, chrome yellow, Naples yellow (true), or flake white, it probably contains free sulphur, or has been insufficiently washed.

In order to test the purity of ultramarine, an easy plan, useful so far as it goes, is this. Boil a small quantity of the sample with distilled water in a wide test-tube for five minutes. Pour the liquid on to a wetted Swedish filter-paper fitted in a funnel. Divide the *clear* filtrate which runs through into 2 parts—to one add a few drops of basic acetate of lead; to the other a few drops of barium chloride solution. No darkening should occur in the first case, nor any white cloudiness in the second. Besides the white adulterants, gypsum and heavy spar, both chessylite and Prussian blue have been found in artificial ultramarine. If chessylite be present, its presence may be detected by warming the sample with ammonia solution, when a blue solution will be obtained. Ultramarine containing Prussian blue acquires a brown hue when warmed with caustic soda solution, while it does not completely lose its blue colour when treated with dilute hydrochloric or sulphuric acid.

When an acid (such as hydrochloric or oxalic) acts upon ultramarine, it disengages both sulphur and sulphuretted hydrogen; a good deal of silica is also then separated in a gelatinous form. The sulphur separated as such may amount to as much as 10 per cent. of the weight of the pigment taken, or it may be less than half a per cent. The sulphur disengaged as sulphuretted hydrogen through the action of an acid varies between half a per cent. and 6 per cent. As these variations have no relation to the depth of colour in the several samples, it is evident that a great proportion of this sulphur does not form an essential part of the blue pigment itself. Other things being equal, it is well to select samples of ultramarine which contain as little as possible of sulphur in any state.

The ultramarines known as 'Guimet's' and 'Heumann's' are of fine quality. Large quantities of this pigment are manufactured in Germany and France, smaller amounts in Belgium and England. The green and lilac ultramarines of commerce are permanent, but not artistically valuable, pigments.

COBALT: *Cobalt Blue—Bleu de Thénard—Kobalt Blau.*

Excluding smalt, which owes its colour to a cobalt silicate, there are at least three pigments which go under the name of 'cobalt' or 'cobalt blue.' The best known of these is a combination of alumina and cobalt oxide; then comes Leithner's or Thénard's blue, a cobalt phosphate on an aluminous base; lastly, there is an aluminous cobalt arseniate very much like the phosphate.

The original and simplest form of cobalt blue, or Wenzel's blue, may be made by calcining strongly an intimate mixture of aluminium hydrate and cobalt oxide. A better way consists in moistening freshly-precipitated aluminium hydrate with a solution of cobalt nitrate, drying and then strongly igniting the mass. It may likewise be prepared by precipitating a solution of sodium aluminate by means of cobalt chloride solution. Thénard's blue may be prepared by mixing about 8 parts of aluminium hydrate with 1 part of cobalt phosphate, both in the moist condition, then drying and strongly calcining the mixture; cobalt arseniate may replace the phosphate. Another variety of Thénard's blue is obtained by adding sodium phosphate solution to a solution of alum containing a little cobalt sulphate. In all the above methods, the complete freedom of the materials used from iron and nickel is essential to the purity and beauty of the blue pigment formed.

The cobalt blues work well in all media. They are unaffected by light, moisture, and oxygen. The best samples of them are practically permanent even in impure air, but ammonium sulphide tends to discolour them. If they appear changed in hue in any oil-paintings, the yellowing of the admixed or overlying oil or varnish must be regarded as the cause. Cobalt blues may be used in fresco-painting; they are unaffected by commixture with other pigments. Cobalt blues do not lose their colour when boiled with alum solution, nor when treated with moderately strong acids.

Cobalt blue as an oil-colour is usually ground with about three-fourths its weight of linseed or poppy oil.

Cobalt blue examined optically is found to reflect much green and violet light as well as blue. Viewed by candle or gas light it acquires a very marked purplish hue.

The introduction of cobalt blue to the palette of the artist may be said to have created a revolution in the style of painting, especially obvious in water-colour landscapes.

CŒRULEUM: *Cœlin — Cerulian — Cerulean Blue — Coelin-Blau — Bleu Céleste.*

When oxide of tin is moistened with cobalt nitrate solution and strongly heated, a greenish blue mass is obtained, which, after powdering and washing, constitues one of the varieties of the pigment known as cœruleum. There are other ways of preparing this substance. One of these consists in precipitating potassium stannate with cobalt chloride, collecting and washing the precipitate, and then mixing it with some pure silica and heating it. A good specimen of cœruleum contained in 100 parts: 49·7 tin binoxide, 18·6 cobalt oxide, and 31·7 silica. Some samples contain calcium sulphate, or lead sulphate, in place of the silica; they are of inferior quality.

Cœruleum is a permanent pigment of a rather greenish-blue colour, without any tendency to the violet cast, so noticeable with other cobalt blues (page 187), when viewed by gas or candle-light. It suffers little, if any, change by exposure to light or impure air, or by commixture with other pigments. It is a sub-opaque, rather earthy pigment, with a moderate tingeing power. Although some painters find it useful, cœruleum may be imitated so nearly by a mixture of ultramarine, viridian, and white that its presence on the palette can easily be dispensed with.

PRUSSIAN BLUE: *Turnbull's Blue—Antwerp Blue—Berlin Blue—Prussiate of Iron—Chinese Blue—Saxon Blue—Bleu de Berlin—Pariser-Blau.*

There are at least three different though closely allied chemical compounds included under the above names, not to mention those varieties of this pigment which contain added or extraneous substances, such as alumina, plaster-of-Paris, or zinc white. The three typical and distinct compounds are:

I. *Soluble Prussian Blue.*—This is made by pouring a solution of ferric chloride or ferric nitrate into an excess of potassium ferrocyanide solution (yellow prussiate of potash), or by pouring ferrous sulphate solution into excess of potassium ferricyanide solution. The blue precipitate formed is washed with distilled water until the wash-water begins to acquire a blue tint. The composition of the pigment thus prepared is, when dry, represented by the formula $K_2Fe_2(CN)_{12}Fe_2$. It contains potassium, and is, in reality, a double ferrocyanide—a 'potassio-ferric ferrocyanide.' It is less stable than either of the other kinds of Prussian blue,

13—2

while its solubility in water causes it to stain the paper on which it is spread in water-colour painting. It should invariably be rejected by artists, although it must be owned that it works very smoothly both in water and in oil. It may always be distinguished from the superior kinds of Prussian blue by very simple tests. One of these consists in roasting a small portion of the dry powdered pigment in a porcelain bason or iron tray, allowing the brown residue to cool, and then throwing it into a little pure water. Then place the mixture on a wetted filter contained in a funnel, and see whether the clear filtrate is alkaline by dipping a piece of yellow turmeric paper into it; if the yellow tint of this paper is reddened, then the Prussian blue belongs to this section. Another test is applied by simply washing some of the powdered blue with warm distilled water on a Swedish filter—the filtrate becomes blue.

II. *Insoluble Prussian Blue* may be prepared by boiling No. I. (the soluble kind) with a solution of ferric chloride, by mixing solutions of ferrocyanic acid and ferric chloride, by pouring potassium ferrocyanide solution into an excess of a solution of ferric chloride, or of ferric nitrate, and heating the mixture for some time, or by precipitating a watery solution of Blue No. I. with an excess of either of the above-named iron salts. It may also be obtained by oxidizing Turnbull's blue (No. III.) with chlorine water or nitric acid. The chemical composition of this pigment is very complex, the simplest empirical formula for it being $Fe_7(CN)_{18}$: it will be seen that it contains no potassium. It always contains some combined water, which cannot be driven off by heat without decomposition of the salt. This blue is more permanent than No. I.

III. *Turnbull's Blue.*—The chief constituent of the original Turnbull's blue (more properly, Gmelin's blue) closely resembles ordinary soluble Prussian blue, and, like it, contains potassium. But the potassium may be removed from it by stannous chloride solution, a substance being produced having the empirical formula $Fe_5(CN)_{12}$, but containing some water. Or the same body may be made by precipitating a solution of ferricyanic acid with a solution of ferrous sulphate or ferrous chloride. This blue is of good colour, but is more difficult to obtain pure than No. II., the other insoluble Prussian blue. Exposed to light, all the forms of Turnbull's blue, pure and impure, have a more decided tendency to become greenish or to fade than No. II.

The ordinary commercial Prussian blue is a mixture, in varying proportions, of the three blues above described. It is made by adding green vitriol (ferrous sulphate) solution to a solution of yellow prussiate of potash (potassium ferrocyanide). The precipitate formed (which varies in colour from a light to a deep blue, according to the amount of ferric salt present in the green vitriol) is then oxidized by means of dilute nitric acid or of a solution of bleaching powder. After having been washed, the substance is treated with hydrochloric acid, and is then again washed with water.

All the above blues are of a very deep blue colour in powder or in the lump, but when pressed or rubbed they all show a coppery lustre. The only one fit for artists' use is the insoluble variety (No. II.), the others being less stable or having other defects. The insoluble form is, moreover, the only one which yields, when roasted, a perfectly satisfactory 'Prussian brown.'

Prussian blue is a transparent colour of great force and richness, and works well in oil as well as in water. In thin washes or layers it has a slightly greenish hue. Its colour is changed by lime and by the weakest alkalies, so that it cannot be employed in fresco or on newly-plastered walls.

Long-continued exposure to strong light weakens and alters the colour of Prussian blue, but the insoluble varieties are less affected than the soluble. When this fading of the pigment in water-colour washes has taken place, a brief sojourn in darkness generally suffices to restore the hue almost to its original depth and quality. This strange phenomenon, which awaits explanation, has been long familiar to artists' colourmen. A sample of Prussian blue (as ordinarily made) in oil, after five years' exposure, had become somewhat greenish, with a loss of about one-tenth of its depth. These changes were more obvious in the pure transparent pigment than in its tint with flake-white. A second specimen, from another maker, similarly exposed, was rather less affected both as to loss of colour and as to change of hue.

Prussian blue was discovered in 1710 by a colour-maker of Berlin, Diesbach by name. When, as is sometimes the case, this pigment is found in water-colour paintings of the seventeenth century, it is scarcely necessary to state that its presence betrays the brush of the restorer or the forger.

Antwerp blue is a sort of Prussian blue lake, the pigment consisting of a colourless base dyed with Prussian blue. According to one method of preparation, a solution of 1 part of green vitriol and 2 parts of zinc sulphate in 40 parts of water is precipitated by adding to it a solution of 4 parts of potassium ferrocyanide in 40 parts of water. The blue colour of the precipitate deepens as it is washed. Alumina is sometimes used as the base upon which the blue pigment is thrown. Antwerp blue is less transparent and less intense in colour than Prussian blue; it has about the same degree of stability.

CYANINE: *Leitch's Blue.*

A mixture of Prussian blue and cobalt blue has been sold under the name of cyanine. It would seem from some recent experiments made with this mixed pigment that it is fairly permanent, even in water-colour painting. It is, of course, not adapted for use in fresco, as the Prussian blue in it is at once turned into rust by the lime of the intonaco. Mixed pigments cannot, however, be recommended, as it is in nearly all cases better for the artist himself to associate together those paints which he wishes to mingle. This plan gives him the opportunity of ascertaining the purity and quality of the several components of his mixtures. In the case of cyanine, it appears that the more permanent constituent, the cobalt, partially protects the Prussian blue from change—a result which is still better seen when a separate wash of cobalt is laid over a wash of Prussian blue. In this case the light which has penetrated through the cobalt particles would seem to have been deprived of those rays which effect the decomposition of the Prussian blue beneath. Such a phenomenon is not unusual, and has been observed in the case of several pairs of pigments having very closely-allied hues.

Indigo.

Indigo has been used either as a pigment or a dye from very early times in India and in Egypt. It is referred to under the name of *indicum* by Pliny; later on the Byzantine writers called it *azorium Romanum*. 'Indigo bagadel'—that is, indigo of Bagdad—is named as early as 1228 in the Marseilles tariffs; in the early English accounts relating to painting works (1274) it is called 'indebas.' It was first largely imported from India into Europe in the seventeenth century by the Dutch.

A large number of different plants yield true indigo. This pigment was once obtained in considerable quantity from a crucifer, Isatis tinctoria, the dyers' weed or woad, the 'pastel' of the French; but the chief source is now

Indigofera tinctoria, a leguminous shrub, probably of Indian, or at least Asiatic, origin. The exports of indigo from India in a single year amount to £4,000,000 worth.

Indigo ($C_{16}H_{10}N_2O_2$) does not exist ready-formed in the plants which yield it, but occurs in the form of a colourless compound, or *glucoside*, which, by combining with water, splits up into a sugar and indigo. It is prepared from the freshly-cut plants, or from the dried foliage, by maceration in water and fermentation, followed by boiling (sometimes lime-water is first added); the dark precipitate which forms is thrown on to cloth-strainers, and finally pressed and dried. The mineral impurities which commercial indigo contains are derived partly from the plant itself, partly from the water used in preparing it, and partly from the lime-water above mentioned; moreover, it is sometimes adulterated. Indigo is easily oxidized by a very large number of substances rich in oxygen, yielding a yellow product called isatine; it is converted into a colourless body ($C_{16}H_{12}N_2O_2$) by many reducing agents.

Indeed, several of the processes of purifying indigo depend upon the reduction of the blue colouring substance, or 'indigotin,' into 'white indigo,' and the subsequent precipitation of the blue matter by exposure to the oxygen of the air. Green vitriol is the commonest reducing agent, and is used in association with lime. The purified indigo prepared by this process, though of fair colour, does not, however, work so well as a paint as the best Bengal indigo treated successively with acid, alkali, acid, and alcohol. Indigotin, if quite pure, has a somewhat purplish cast in thick water-colour washes; this hue is observable with this substance whether obtained by sublimation or by Fritzche's process with grape-sugar, caustic soda, and alcohol.

The impurities in commercial indigo constitute from 20 up to 70 or 80 per cent. of its total weight—the average is about 50. They consist mainly of mineral matter, indigo red, indigo brown, and nitrogenous compounds. Much of the mineral matter may be removed by digestion in hydrochloric acid, followed by treatment with boiling water. Sodium hydrate solution dissolves the indigo brown, while strong alcohol takes away the indigo red, which amounts to nearly 4 parts in 100 of the original indigo. After treatment with these three solvents, the residual purified indigo is of an intense and very beautiful hue. Java indigo is generally of very good quality, that from Bengal comes next, and then the indigo from Oudh, Kurpah, and Madras. Japanese indigo is generally poor.

Indigo frequently receives no purifying treatment previous to its being ground into a fine powder suitable for admixture with oil or with gum and the other media of water-colours. The necessity of choosing the purest and finest samples of the commercial dye-stuff is of course evident, but it is better in every case to adopt the processes of purification named in the preceding paragraph. No sample of purified indigo should leave, after being burnt, more than 3 per cent. of ash.

This rich and transparent blue is, unfortunately, gradually oxidized and browned when exposed to light. In thin washes of water-colour it disappears rapidly in the sun's rays, much more slowly when submitted to diffused daylight. The following figures approximately represent the reduction in force of indigo as a water-colour when exposed to sunlight :

Original intensity	10
After two years	1
After ten years	0

Indigo in cake is sometimes less affected by sunlight than the moist preparations. As an oil-colour, indigo loses from one-third to one-half of its intensity when exposed to sunlight for five years, its hue being at the same time altered, in different specimens, either to a greyish or a greenish blue; the change is more conspicuous when the indigo has been mixed in tint with flake or other white. Locked up in copal or amber varnish it is more slowly changed. The fading is due to oxidation.

Indigo may be replaced advantageously by ultramarine or by a good Prussian blue, either being associated with a little ivory-black.

Several pigments, such as aureolin, true Naples yellow, and all the chromates, have a very marked effect upon indigo.

In order to ascertain whether the fading of indigo as a water-colour, on exposure to sunshine, was increased by the presence of alum in the paper, a series of comparative experiments were made. A pale tint of indigo was spread upon (1) paper free from alum; (2) paper washed with alum solution; (3) paper containing a trace of alum; (4) paper which had been washed with weak ammonia-water after having received an alum-size. After six months' (April to September inclusive) exposure to sunlight, all the four specimens showed complete extinction of the blue pigment, the disappearance of the colour from No. 2 having, however, been a trifle more rapid than in the other cases. Honey and glycerin, owing to their hygroscropic character, appear to hasten the fading, so does water-vapour; for a *dried* slip of indigo-washed paper sealed up in a glass tube loses its colour less quickly than one in its ordinary moist condition, when both are exposed side by side to sunshine. A tightly-framed water-colour drawing presents, of course, a close analogy to the second or more unfavourable set of conditions.

The effect, if any, of the several iron reds, such as light red, Venetian red, and Indian red, in accelerating the fading of indigo in sunshine has been tested in many ways. Provided the three red pigments above-named be free from soluble salts, and especially from sulphates, they are equally innocuous; the indigo disappears with or without them at the same rate. In one instance only a wash of Venetian red with indigo lost its indigo completely three weeks before the destruction of the blue in the parallel experiments was complete, which was the case in six months. It is impossible to assert that this result was not due to a smaller amount of indigo having been present in this particular case, but I believe it to have been due to this sample of Venetian red having been a spurious one prepared from colcothar.

In comparing under the same conditions the relative stability, when exposed to light, of different indigoes, it appears that 'Bengal refined' is superior to the best 'raw Bengal,' and even to the indigotins obtained by the green vitriol and glucose processes.

The superiority, when exposed to light, of refined indigo —that is, of indigo which has been purified by the treatment with acid and the other solvents already named—although a result of recent experiments, seems to have been ascertained in the time of De Mayerne. One of the authorities (Elias Feltz, of Constance) quoted by De Mayerne affirms that this pigment may be rendered safe by steeping it in vinegar and exposing it to the sun for several days; the vinegar is then to be poured off, and the paste when dry ground in oil. Another plan recommended (in a note, dated 1642) by Feltz consists in grinding the indigo with a little calcined alum; this plan is, however, objectionable for several reasons. The treatment of the indigo with the solvents previously named, which are adapted to remove its natural impurities, is the only legitimate plan. It is scarcely necessary to add that there is abundant evidence to prove that in De Mayerne's day indigo was generally regarded as an unstable colour. Under all circumstances no indigo should be employed in painting unless it be completely free from acids.

Although it is not possible to ascertain the exact richness of commercial indigoes by any tests save those in use by professional chemists, yet it is easy to learn a good deal about any particular sample by the application of a few simple experiments. A good sample should appear homogeneous, and should float on water. Dried at 100° C., it should not lose more than 6 parts per 100 of moisture. When rubbed with a hard smooth substance it should show a coppery lustre. One hundred grains when burnt should not leave more than 10 grains of grey ash. It should dissolve perfectly in four times its weight of fuming sulphuric acid. Starch, gypsum, clay, chalk, steatite, and Prussian blue are amongst the adulterants of indigo.

Indigo dissolved in four times its weight of fuming oil of vitriol (kept cool) forms a liquid from which, after slight dilution and filtration through asbestos cloth, potassium carbonate precipitates a fine blue compound, which has been used in water-colour painting under the name of indigo carmine. It is not a safe pigment.

In examining old water-colour drawings, it will often be found that the parts protected by the mount or frame show the indigo used in compounding the greys of clouds and the greens of vegetation perfectly intact, while it has completely left the exposed parts. In many cases where indigo is supposed to have withstood long exposure to light, it will be found that the blue used has been Prussian blue, modified by admixture with other pigments. But in many old pictures and drawings it will be found that the unnatural bluish hue of the foliage represented is due to the complete loss of rapidly evanescent yellows, rather than to the entire stability of the indigo with which the greens have been compounded. Nor must it be forgotten that indigo was not the only blue in such mixed greens, blue carbonate of copper and even lapis-lazuli having been extensively employed at that time. The latter pigment is perfectly permanent except in the presence of acids.

SMALT: *Royal Blue—Dumont's Blue—Zaffre—Bleu de Smalte—Smalte.*

Glass and vitreous mixtures containing cobalt, and of a rich blue colour, have been known for ages, although copper, not cobalt, was the colouring principle of most of the antique materials of this class. The smalt now made is a very deep-blue glass (appearing black in the lump), consisting essentially of cobalt and potassium silicate. It generally contains, in 100 parts, silica, 65 to 71; potash, 16 to 21; and cobalt oxide, 6 to 7 parts: a little alumina is always present. In the inferior varieties the oxides of iron and nickel always occur.

In order to prepare a pigment from the dark cobalt glass, it is fused, and then poured into cold water. After the disintegration thus effected, the glass is ground into a moderately fine powder, and submitted to the process called elutriation, or washing-over. The finest particles, which take longest to settle from the wash-waters, are the palest in colour; the larger particles, though of richer hue, are very difficult to use as a paint.

Smalt is rarely employed now as an artists' pigment, cobalt blue and artificial ultramarine having been very advantageously substituted for it. For not only is it a difficult pigment with which to work, in both water and oil, but it is gradually altered by moisture and by the carbonic acid of the air, becoming paler and greyer; moreover, the finer the state of division in which it exists, the more rapid is the change. It will be found that even spring or distilled water is competent to start the decomposition of smalt. For if a little of this pigment be placed on a piece of yellow turmeric-paper, and moistened with clean water, there will soon be formed a red stain beneath the smalt—proof of the liberation of some of the alkaline constituent (potash) of this blue glass.

CHESSYLITE: *Blue Verditer—Bice—Mountain Blue—Azurite—Cendres Bleus—Bergblau.*

This copper mineral differs from malachite in containing less hydrate, or more carbonate of the metal, its composition being represented by the formula $2CuCO_3, CuH_2O_2$. Its best-known locality is Chessy, near Lyon, in France, but very fine specimens are found at Wallaroo and Burra-Burra, in South Australia, and in the district of Perm, Siberia.

This blue pigment has been prepared artificially, but the natural substance is far less liable to change on exposure to impure air. The introduction of cobalt blue, and more particularly of artificial ultramarine, has practically caused the disuse of blue verditer.

Many old pictures, especially those in tempera, afford evidence of blue verditer having become green. With the extensive range of more permanent blues at the command of the modern artist, there is no need to retain this pigment on the palette.

CHAPTER XVIII

BROWN PIGMENTS

RAW UMBER: *Levant Umber—Terre d'Ombre—Umbraun—Umbra—Kölnische Erde—Terra Ombra.*

THIS earth is found in several localities; the best variety has come for some time past from Cyprus. A considerable number of Cypriote specimens, of several nuances, some excellent, were shown in the Colonial Indian Exhibition of 1886. It differs chemically from the yellow and brown ochres in several particulars, notably in the presence of a considerable quantity of one of the higher oxides of manganese (Mn_3O_4, or MnO_2), and in the small proportion of water which it contains. Samples from English localities are poor in iron; one Derbyshire specimen gave Mr. G. H. Hurst no less than 30 per cent. of barium sulphate. A recent analysis of my own, made with a choice sample from Cyprus, showed the following percentages:

Water, given off at 100° C.	4·8
Water, given off at a red heat*	8·8
Iron oxide (Fe_2O_3)	48·5
Manganese dioxide (MnO_2)	19·0
Lime	1·4
Magnesia	0·5

* Includes a little organic or bituminous matter.

Alumina	2·1
Phosphoric acid (P_2O_5)	0·9
Silica	13·7
Carbonic acid, etc.	0·3

This sample had the peculiar greenish hue so much prized by artists. It should be stated that a part of the manganese probably existed as Mn_3O_4.

Before being used as a paint, this brown mineral is finely ground, washed with water, and then dried at 100° C., or at a slightly higher temperature. When, by a stronger heat, the whole of its water has been expelled, the umber acquires a reddish hue, and is then the pigment known as *burnt* umber. This change of colour is due to the passage of the brown ferric hydrate into the red ferric oxide, and to an increase in the proportion of the red-brown manganese oxide present.

Raw umber in powder, after having been purified, soon acquires a very slight reddish hue on exposure to light and air; it is a good plan to place the undried, finely-ground mineral on trays in the sunshine before completing its desiccation and mixing it with oil or other medium.

Raw umber is permanent when used with each or any of the painting media: the slight yellowish or dull aspect which it acquires in oil may be traced to the augmented translucency of the paint, and to the yellowing of the associated oil. Umber is without action on other pigments. A very few samples of umber, used as a water-colour, have been observed to fade slightly after from five to ten years' exposure to sunlight. But this deterioration is due to the presence of traces of brown peaty acids, or 'humus' substances, which occasionally occur in the umbers from certain localities.

Raw umber possesses a semi-opaque, citrine-brown colour;

it works and dries well in oil. Associated with transparent blues, it yields soft, quiet green hues; it is invaluable both in figure and landscape painting. As a priming or first painting-ground, it is apt, like most dark pigments, to become more conspicuous in time, owing chiefly to the translucency which the superimposed painting gradually acquires.

Raw umber is not subject to adulteration, but a ferruginous brown-coal has been occasionally substituted for the true mineral. The great variation in quality shown by the umbers of commerce is due, in great part, to the difficulty of securing, even from the same mine, continuous supplies of the same excellence.

Burnt Umber.

It has before been pointed out that raw umber, from which burnt umber is prepared by calcination, is not an ordinary ochre, but owes its colour in great measure to the presence of a considerable amount of some compound of manganese. The exact constitution of raw umber is not, however, known, although the very slight change in hue which occurs when it is roasted negatives the idea that it contains any considerable proportion of the ordinary ferric hydrates. If these were present in notable proportions, roasting would certainly redden raw umber much more than is actually the case.

Burnt umber differs in quality and hue from the raw earth mainly in being more translucent, and of a warmer brown. Although some Continental authorities affirm that this pigment darkens and becomes purplish in course of time, I cannot regard it as otherwise than perfectly permanent, and as exerting no action on stable pigments.

RAW SIENNA: *Terre de Sienne—Siena-erde—Terra di Siena.*

This earth, which is found chiefly in Tuscany and the Hartz, is a particularly rich variety of yellow ochre, and contains a large proportion of a ferric hydrate. Mr. G. H. Hurst ('Chem. News,' vol. lix., p. 172) gives three analyses of raw sienna. In these the range in the percentages of the more important constituents is as follows:

Hygroscopic water	8·2 to 17·5
Combined water*	9·0 „ 12·4
Manganese dioxide...	0·6 „ 1·5
Iron oxide (Fe_2O_3)	45·8 „ 59·7
Silica	5·0 „ 17·4

Raw sienna is prepared for use by crushing, sifting to remove sand, grinding and washing in the same way as in the case of yellow ochre. It should be noted that when lumps of fresh raw sienna are first broken and exposed to the air, their surfaces acquire a slight olive-green nuance. In order to avoid any disadvantage which might arise from such an alteration taking place subsequently, it is a good plan to expose the crushed earth to the air and light previous to its final preparation as a pigment. Before grinding it in oil, it should be cautiously dried at 60° C. to remove the greater part of the accidental or hygroscopic moisture.

Raw sienna is somewhat deeper in tint and of a warmer and browner hue than yellow ochre. Owing to a trace of organic or peaty matter which it contains, raw sienna is liable to become rather less brown and more yellow by long-continued exposure to strong light. It is without action on other pigments, and is available for water-colour, oil, fresco, and tempera painting.

* Includes traces of organic matter.

Raw sienna is not subject to adulteration, but it is well to be certain that the pigment has been well washed.

Burnt Sienna.

The roasting or calcination of raw sienna produces a very great change in its hue as well as in the depth of its colour. The ferric hydrate of the raw earth becomes wholly converted into ferric oxide, this change being accompanied by a great increase in the translucency and depth of the colour.

When small fragments (or the coarse powder) of raw sienna are calcined, it will be noticed that the change of hue which occurs is not quite uniform, some points being of a brighter and redder brown than others. A very fine powder, thoroughly stirred during roasting, shows this peculiarity in a much less marked degree.

Burnt sienna possesses a very beautiful, warm, reddish-brown hue, which cannot be exactly imitated in translucency and depth by mixtures of other pigments. It is permanent, without action on other pigments, and not liable to adulteration. It is available for use in every method of painting. Some fine and permanent foliage-greens may be made by associating viridian with burnt sienna.

Caledonian Brown.

This brown, although a natural earth, presents very much the appearance of burnt sienna. It contains a small quantity of combined moisture. It consists mainly of the brown hydrates and oxides of manganese and iron. When calcined it loses its ruddy hue and becomes almost black—a black with a slight brownish hue.

Caledonian brown, whether raw or burnt, is a permanent and innocuous pigment, which is well adapted for oil and tempera painting.

Vandyke Brown.

Three brown pigments pass in commerce under the name of Vandyke brown. The first is made by calcining certain very ferruginous earths or brown ochres; the second is nothing more than a dark-brown variety of colcothar; the third is a kind of brown earth containing, along with some iron oxide and hydrate, a good deal of organic substance in the form of humus or bituminous matter. The first and second kinds are permanent and innocuous, but the third kind will not resist the prolonged action of light, becoming paler and redder in the course of time. The discrepancies in the published statements as to the permanence of this pigment are thus readily explained: it is to be regretted that most of the samples of Vandyke brown now met with in England belong to the third kind, and therefore fade quickly in water-colour, more slowly in oil. This sort may be recognised by the dark sublimate which it yields when its powder is heated in a test-tube, as well as by the change in colour and great loss of weight which it then shows. We shall designate this less satisfactory variety in the Tables of Permanent and Fugitive Pigments in the present volume as Vandyke brown B., the other kinds being called Vandyke brown A.

Cologne or Cullen earth, and Cassel brown or Cassel earth, are soft, impure varieties of brown coal or lignite. They vary in fixity, some of them being even more easily bleached by light than Vandyke brown B.; these should not find a place on the palette of the artist. When slightly roasted, a part of the brown organic matter in these earths is charred or carbonized, and the substance becomes darker, duller, and decidedly less alterable by exposure. Some of the so-called Cologne earth now sold is merely Vandyke brown B., slightly changed by gentle roasting; it is then rather less alterable. We have met with some specimens of Cassel earth which proved practically permanent in oil, but even these faded quite distinctly when exposed to strong light after having been mixed with flake-white.

CAPPAGH BROWN: *Euchrome—Mineral Brown.*

This earth was found on the estate of Lord Audley, in the Cappagh Mine, which was opened in the year 1814, and is situated about ten miles west of the town of Skibbereen, in the county of Cork. In composition and general characters it resembles raw umber, but has a more reddish hue. It contains ferric hydrate and ferric oxide, with a considerable amount of one of the oxides or hydrates of manganese. It gives off a good deal of water when heated to 100° C., and acquires a rich reddish-brown colour, not unlike that of burnt sienna, and almost identical with that of Caledonian brown. The specimens examined contained mere traces of organic matter, so that it cannot be regarded as a kind of bog-earth or peat, although it is, of course, possible that there may be another mineral found in the same locality which might be correctly so designated. The following analysis represents the composition in 100 parts of a characteristic specimen of this pigment:

Water, given off at 100° C.	18·7
Water, given off at a red heat	11·6
Ferric oxide (Fe_2O_3)	34·4
Manganese dioxide (MnO_2)	27·2
Alumina	2·6
Lime	1·1
Magnesia	*trace*
Silica	4·6
Phosphoric acid (P_2O_5)	0·4

It is probable that a part of the manganese really exists in the form of the red oxide (Mn_3O_4). The large quantity of water present in this mineral in a loosely-attached form (hygroscopic), amounting to nearly one-fifth of the weight of the pigment, indicates the desirability of cautiously drying the substance previous to grinding it in oil. A temperature of 60° C. should not be exceeded.

Cappagh brown works well in oil, particularly if it be dried at a heat below that of boiling water before it be ground in oil. It is an innocuous pigment, but its permanence has scarcely been sufficiently tested. A rub of Cappagh brown in oil, exposed to strong sunlight for one month, lost a little of its yellow, assuming a somewhat ruddier hue; the change, however, was very slight, and did not appear to have increased after continued exposure for five months more.

BISTRE: *Bister—Brauner Lack—Russbraun.*

Bistre is prepared from the tarry soot of certain woods, especially from that of beech-wood, by the following process: The soot is finely ground and sifted, and then the powder is digested with successive portions of hot water until the latter no longer acquires a brown or yellow tint; the washings are then evaporated with suitable quantities of gum-water and glycerin, and the residue preserved in the moist state. To form cake-bistre the glycerin is omitted, but more gum is employed. Bistre is not used as an oil-colour.

The tarry matter in bistre is its element of weakness. By exposure to strong light this tarry matter oxidizes, and the residual pigment becomes cooler in hue and paler. Professor Ogden Rood, experimenting with a weak water-colour wash of bistre, found that it lost 19 parts of its original intensity of 25 by an exposure of three and a half months to the summer sun. I have not found so considerable a change to occur with the samples of bistre which I have tried; but this pigment varies much in composition, being obtained from the soot of different woods as well as from that of peat. The most fugitive preparation is that made by extracting the vegetable tars referred to above with water containing a little ammonia, and then precipitating the deep-brown liquor with a slight excess of acetic acid.

ASPHALTUM: *Bitumen—Mineral Pitch—Antwerp Brown—Mummy—Asphalt Braun.*

Asphalt, asphaltum, or mineral pitch, has been long used as a pigment. The best known is that from the Dead Sea (*Lacus asphaltites*). Other abundant sources of this carbonaceous mineral occur in Trinidad; Caxitambo and Berengela, in Peru; Val de Travers, Neufchâtel; Avlona, in Albania, etc.

Asphalt is rather a mixture of minerals than a single mineral; it is, moreover, very variable in the nature, character, and proportion of its constituents. Essentially it consists of a number of liquid, semi-solid and solid, colourless hydrocarbons (related to the paraffins), associated with certain ill understood dark-brown or black substances, which constitute the useful part of the raw material. The best varieties for artistic use are those which contain the smallest proportion of the above-described hydrocarbons, for to the presence of these the treacherous character of asphalt as a pigment is due. On this account the hardest, most earthy, and most brittle kinds should be chosen, and the crushed samples should always be submitted to a temperature of at least 250° C. before being ground in oil or turpentine. The operation of roasting native asphalt—keeping it over a slow fire 'till it will boil no more and becomes nearly a cinder'—was recommended by Williams in his 'Essay on the

Mechanic of Oil-Colours' (1787), and furnishes a perfectly satisfactory and safe product.

If carefully-selected asphalt be submitted to either of the processes named above, and then be moistened with spirits of turpentine, and ground in drying-oil (prepared with borate of manganese), a paint is obtained which neither cracks nor moves on the canvas like the unpurified material. Its fixity is further ensured by mixing it with a little copal varnish, and more particularly by associating it with a denser pigment, such as umber or flake white. It is superior to the imitative asphalts made from coal-tar, now largely sold in lieu of the original and genuine product. The disadvantages attending the use of these coal-tar browns and of ordinary asphalt are twofold. Not only are they treacherous on account of their easy fusibility, but they are liable to *stain* contiguous pigments by reason of their solubility in oil or varnish. When used successfully by the older artists they were always introduced sparingly, or were largely commingled with more solid paints.

'Mummy,' as a pigment, is inferior to prepared, but superior to raw, asphalt, inasmuch as it has been submitted to a considerable degree of heat, and has thereby lost some of its volatile hydrocarbons. Moreover, it is usual to grind up the bones and other parts of the mummy together, so that the resulting powder has more solidity and is less fusible than the asphalt alone would be. A London colourman informs me that one Egyptian mummy furnishes sufficient material to satisfy the demands of his customers for seven years. It is perhaps scarcely necessary to add that some samples of the pigment sold as 'mummy' are spurious. Mummy was certainly used as an oil-paint at least as early as the close of the sixteenth century.

Asphalt, after having been heated to drive off the hydrocarbons previously alluded to, cedes to ammonia a considerable quantity of a dark-brown colouring-matter, which in this way may be made available for water-colour painting. The ammoniacal solution is either evaporated slowly, to the consistence of a thick syrup, after the addition of a little gum and glycerin, or it is precipitated with acetic acid, and the precipitate (after washing) is mixed with gum and glycerin, and then partially dried until it has acquired a suitable degree of consistency. But the water-colour paint thus made is not permanent.

Merimée's process for preparing asphalt for use as an oil-colour cannot be recommended. He introduces shellac, white wax, and Venice turpentine into the mixture, as well as a large proportion of boiled linseed oil. This preparation constitutes a very treacherous pigment.

PRUSSIAN BROWN.

This pigment, as usually met with in commerce, contains a considerable quantity of a soluble salt of potash, and is not fitted for the use of artists. But this impurity need not be present, as it may be removed by a thorough washing of the powdered colour with boiling water. A still better plan is to prepare Prussian brown from one of those varieties of Prussian blue which contain no potassium. It is made by throwing small pieces (the size of hazel-nuts) of Prussian blue upon a plate of iron maintained at a red heat. Each fragment burns like tinder, and, if care be taken to employ the right heat for the proper time, will show a mixed hue, partly yellowish brown, partly reddish brown, and partly black. The product, still hot, is thrown into water, ground, washed, and dried. Prussian brown thus made has a rich colour of considerable translucency and good drying character. In hue it is warmer than asphalt. It is quite permanent.

CHAPTER XIX

BLACK PIGMENTS

INDIAN INK: *Chinese Ink—Japanese Ink—Encre de Chine—Chinesische Tusche.*

THIS ink has been prepared in China for at least 2,000 years. It consists essentially of a very fine lamp-black, associated with gelatin, and scented with musk, camphor, cloves, or rose-water. The lamp-black employed is derived from the imperfect combustion of oil or of pine-wood. The oils chiefly used are those of *Sesamum indicum*, *Cannabis sativa*, and *Dryandria cordata*; but in some factories rape-oil, bean-oil, or the oil of *Gleditschia sinensis*, is employed. According to the treatise of Chen-ki-suen, which was written A.D. 1398, these oils are burnt in small earthenware lamps in the presence of a limited supply of moist air. The smoke is collected in earthenware conical covers; from these the condensed soot is removed at short intervals, care being taken to preserve those portions only which are free from tarry products. The soot is finally sifted, and reduced to an extremely fine powder. Lac-resin, rock-oil, as well as many kinds of wood, have been employed for the preparation of this carbonaceous basis of China ink in different parts of the empire, and at different times. It would appear that from all of these combustibles, if due care be taken, an excellent product may be obtained. The next step in the manufacture consists in the incorporation of the soot, prepared as above described, with the necessary amount of size. The size employed should be made from a mixture of clean fish-glue and parchment-size; this is thoroughly mixed, while warm, with the fine soot. The paste thus formed is made into balls, which are heated, for some time, at the temperature of boiling water, and then fashioned roughly into sticks. These are subjected to repeated blows —some hundreds, at least—from a hammer; they are re-heated occasionally during this operation, in order to prevent them from becoming hard. The perfume, consisting of musk and camphor, mixed with a little rose-water, is incorporated with the mass. The material, after further pounding and beating, may then be pressed into the wooden moulds which are to give it its final form. A tedious process of drying follows next, the sticks of ink being finally packed in the ash of rice-straw frequently renewed. After the removal of adhering ash, the sticks are cleaned, gilt or otherwise ornamented, and polished with an oily brush.

If the soot used in the manufacture of this ink be contaminated with tarry or empyreumatic matters, it yields a brownish-black product; the purer and the finer the state of division of the carbon, the more intense is the black, and the more neutral are the greys which it yields on dilution with water, or by admixture with opaque white pigments.

There are many qualities of this ink made in China. The best kinds are hard and homogeneous; they show a lustrous black fracture, and, when rubbed with water, do not give rise to the separation of any curdy particles. Those having, in tint with water, a bluish or violet tinge are the most esteemed; the pure black come next, and the brownish or yellowish black last.

A stick of Indian ink should be rubbed on the palette, in straight lines, backwards and forwards. A very convenient ink-slab for this purpose forms part of the fittings of the Japanese writing-case or box, called 'suzuri-bako.' It consists of an oblong block of dark slate, in which a sloping cavity has been hollowed out. The slight 'tooth,' or roughness of the surface, greatly helps the rubbing-down of the ink. A strong wash of this ink, on sized paper, should not be affected, when once dry, by having water brushed freely over it. Bluish-black and pure black Indian ink is permanent under prolonged exposure to sunlight or impure air; but the brownish varieties become paler, less brown, more neutral, by the action of light. Like most carbonaceous matters, Indian ink occasionally tends to absorb to a small extent some of the organic pigments which may be brought into contact with it in the process of colouring a drawing.

Indian ink is available for tempera-painting, but cannot be used with oil.

Indian ink has been imitated with fair success in France. The chief difficulty in preparing it seems to lie in the preparation of a suitable gelatinous medium with which to mix the purified lamp-black. A size prepared from washed fish-glue (by partial precipitation with tannin, and subsequent solution of the precipitate in more size) affords a good binding material. The size, in true Chinese ink, forms a very large part of the total weight, varying from one-fourth to over one-third; but the European imitations rarely contain as much. Japanese ink is generally inferior to Chinese.

Chinese amateurs sometimes form collections of rare and old kinds of Chinese ink, occasionally giving for sticks bearing the marks of renowned makers not less than their weight in gold. Such specimens are prized for their rarity, not on account of any special merits as pigments which they may possess; and a Chinese ink collector regards it as an unpardonable offence to moisten any of his specimens with a view to testing their quality.

LAMP-BLACK: *Noir de Lampe—Noir de Fumée—Noir de Houille—Russ.*

When resins, resinous woods, fatty oils and fats, paraffin and paraffin-oil, or coal-tar oils, are burnt with an insufficient supply of air, a considerable part of the carbon they contain may be deposited in the form of soot. This soot is not, however, pure carbon, but retains variable proportions of the tarry products of imperfect combustion or destructive distillation; these impart to lamp-black a more or less pronounced warm brownish hue, except in the cases in which it has been prepared by processes specially devised to intercept the tarry and oily products in question. Sometimes small furnaces, sometimes large lamps with long wicks are employed in its manufacture, the soot given off being collected in two or more receivers; the soot first deposited contains the larger part of the impurities. This point may be illustrated by the simple experiment of depressing a white porcelain plate into the flame of a candle; the nearer the plate to the wick, the browner will be the soot deposited.

Lamp-black carefully made is an unalterable pigment; but its employment for pictorial purposes has frequently been objected to by writers on artistic practice as tending to heaviness and opacity in the shadows. With the exception of a few pigments of organic origin, which in water-colour painting have a tendency to cede some of their colouring matter to any kind of carbonaceous black, lamp-black exerts no injurious influence upon any pigments which are stable when used alone.

Numerous carbonaceous substances have been used as black or brown-black pigments. Amongst these anthracite and common bituminous coal may be named. Van Mander and De Mayerne both mention forge or pit coal.

CHARCOAL-BLACK: *Blue-Black—Vine-Black—Frankfort-Black—Noir de Vigne.*

When non-resinous woods and woody tissues are strongly heated in crucibles or other almost completely-closed vessels, the residue contains the greater part of the carbon of the original material, and preserves its form. As the charcoals thus made contain some soluble mineral matter, particularly potash salts, they should be coarsely ground and thoroughly washed with hot water previous to their being finally converted into paints. They are also hygroscopic, and therefore need to be dried before being ground in oil.

Amongst the best materials for producing these charcoal or vegetable blacks may be named the hard shells or stones of plums, almonds, and similar fruits; coco-nut shell broken into small pieces; cork cuttings, and the twigs of beech and shoots of vine. Coco-nut yields the densest, most velvet-like black; vine-shoots a less solid black, having a bluish hue. Washed wine-lees yield a rich black—Frankfort-black.

When charcoal has been prepared at a sufficiently high temperature, and then has been thoroughly washed, there is no question as to its permanence as a pigment in all media. But if it retains tarry matters, or has been imperfectly carbonized, then it may become greyer or less brown (when used in thin washes) after long exposure to light. This change is due to the oxidation of the tarry matters, or of the brown substances which are intermediate in composition between the original vegetable materials and the carbon, which is the final product of their torrefaction out of contact with the air. And one peculiar property, possessed in varying degrees by charcoal of every kind, must not be lost sight of. Charcoal withdraws the majority of organic colouring matters from suspension, and even from solution in water. A pale tint of rose-madder in water, if a pinch of coco-nut or vine-black be added, becomes rapidly decolourised, the pigment being completely absorbed, although not really destroyed. Animal charcoal (bone-black, ivory-black) exerts a still more energetic action of this character; but lamp-black is less powerful.

It is to be regretted that the beautiful bluish blacks, derived from such materials as the shoots of the vine and beech, are frequently fraudulently imitated by mixtures of lamp-black and indigo. The purple vapours of indigotin, given off when such a mixture is heated in a test-tube, betray the adulteration at once.

IVORY BLACK: *Noir d'Ivoire—Elfenbeinschwartz.*

Waste ivory, in the form of turnings, shavings, and saw-dust, if charred in closed vessels, leaves a black residue, which, however, consists essentially of bone-earth (calcium carbonato-phosphate), stained by the presence of a few per cents. of carbon. On exposure to the air, ivory black absorbs, not only the gases of the atmosphere, but also a very considerable proportion of moisture. On this account ivory black, just previous to its being ground in oil, should always be thoroughly dried. As this substance has a strong decolourizing power when placed in contact with moist organic pigments, it is better adapted for use in oil painting than with water colours. This decolourizing property belongs also to the animal blacks obtained by charring horn and bone shavings—products which are frequently substituted for genuine ivory black. Sir Charles Eastlake tells us that in a collection of specimens of water-colour tints inserted in the Mayerne manuscript the 'hartshorn' black, made from *cornu cervinum*, remains very intense.

Ivory and bone black are perfectly permanent black pigments; the latter is quite free from the bluish tinge which characterizes most of the vegetable charcoals. Both are better adapted for oil than for water-colour painting, for which vine black is preferable.

Black Lead: *Plumbago—Graphite.*

The material of which so-called *black-lead* pencils are made is essentially nothing but one of the three forms in which the element carbon occurs; diamond is another; lamp black may be taken as a good representative of the third form. Graphite is the most appropriate name for the mineral commonly known as black lead, but which in reality contains no lead, and is in no way related to that metal. Graphite was found of good quality at Borrowdale in Cumberland, a locality which has been stated, probably erroneously, to be now exhausted; but excellent supplies of this mineral come from the Albert Mine, Siberia, Mount Alibert in the Ural, Ceylon, and many other places. Graphite, however, is never found in a state of perfect purity, always leaving, when burnt, some incombustible matter or ash.

In order to prepare graphite for use as a pigment, the purest pieces should be selected; these should be at first coarsely ground, and then heated with strong hydrochloric acid to remove iron and alumina. After washing, the thus-far purified material should be placed in a leaden or platinum vessel and gently warmed with a solution of hydrofluoric acid to remove silica. The graphite, after this treatment, is to be washed with abundance of water, and ground to a very fine powder, which is then ready for admixture with gum water or, after drying, with oil.

Graphite has a very dark-grey colour, and forms pure tints of grey with white pigments. It is equally well adapted for use in water-colour, oil, fresco, and tempera painting. It is absolutely permanent, and without action on other pigments.

Graphite is sometimes adulterated with charcoal or lamp black. When a very thin layer of the suspected pigment is spread on a piece of platinum foil and strongly heated over a lamp, the lamp black will burn away, leaving the true graphite unaltered. If a change of hue or tint occurs, this is a sign of the presence of the above-named adulterant.

SEPIA.

The dark-brown colouring-matter from the ink-bag of *Sepia officinalis*, *Loligo tunicata*, and other species of cuttle-fish common in the Mediterranean and Adriatic, has not been thoroughly investigated from a chemical point of view. The pure pigment, which constitutes four-fifths of the weight of the dried ink-bags as they occur in commerce, partakes of the character of a weak organic acid, and is soluble in alkalies and precipitated by acids. In preparing the substance for artists' use, it is commonly first removed from the sacs containing it, dissolved in soda or ammonia solution, and then, after straining the solution, thrown down by neutralizing the alkali with hydrochloric acid; the precipitate is then washed by decantation, and dried. Sometimes the filtered ammoniacal solution is used directly as a liquid ink. The chief impurities of the natural sepia are salts of lime and magnesia, which may be partly removed by washing the dried and crushed sepia with dilute hydrochloric acid previous to dissolving it.

Sepia is of a redder or warmer brown hue than bistre, but it is not so reddish as Vandyke brown. Of these three pigments, there is no question that sepia is the least alterable. It is not, however, permanent when exposed to sunshine, although in darkness or in diffused daylight it suffers no appreciable change either in depth or quality. It might, indeed, have been supposed that sepia would prove unalterable, from the consideration of the observed fact of the pigment from the ink-bags of fossil cuttle-fish showing now, when used as a water colour, the same hue and the same chemical characters as recent sepia from the Mediterranean cuttle-fish of to-day. But it must be recollected that this fossil pigment has been excluded completely from the adverse influence of light, and in great measure from that of air, during the long ages which have elapsed since it was embedded in rock. And it is light which, in the case of the majority of organic pigments, is the determining cause of the changes brought about by the combined presence of moisture and atmospheric oxygen.

Sepia is not employed as an oil colour.

CHAPTER XX

CLASSIFICATION OF PIGMENTS

THE painter naturally classes pigments according to their colour, or more exactly according to what are called the *constants* of colour, namely—hue, brightness and purity. He also takes into account transparency and opacity, although these terms are conventional and comparative only, since no pigment is perfectly transparent, none perfectly opaque. Another basis of classification, and a very important one, that of stability, will be considered in the next chapter; at present we are concerned with none of these methods of grouping pigments, but with others founded either upon their origin, or their physical characters, or their chemical composition.

Pigments are often classified into two groups, the mineral, and the organic. It is necessary to divide these groups further, in some such way as this :

MINERAL PIGMENTS.	Natural : as ochre, terre verte, ultramarine.
	Artificial : as aureolin, viridian, cobalt blue.
ORGANIC PIGMENTS.	Animal : as Indian yellow, carmine, sepia.
	Vegetable : as gamboge, sap green, indigo.
	Artificial : as Prussian blue.

Such a classification brings into prominence one marked distinction between the two groups, for, in accordance with one's expectation, the mineral pigments are, as a rule, characterized by a much higher degree of permanence than those of organic origin. [The chemist looks upon all compounds containing carbon, save the carbonates, as organic; but at the same time the distinction between organic and inorganic, or organic and mineral, is nothing more than a convenient convention.]

Other bases of classification are afforded by physical or mechanical characters. Pigments may be fixed or volatile, soluble or insoluble, crystalline or amorphous. There are difficulties in carrying out these schemes of classification, and it will be found that distinctions of physical character are utilized to the best purpose when connected with such a chemical classification as is offered below.

The simplest chemical classification is this:

ELEMENTS; as graphite, lamp-black, gold.
COMPOUNDS; as aureolin, viridian, vermilion.
MIXTURES; as yellow ochre, brown pink, rose madder.

After considerable expansion and some rearrangement, the differences just indicated afford a reasonable basis for a chemical classification which will prove of real service in judging of the degrees of stability, and of the possible interaction of pigments. The mixed and even indefinite character of many pigments, and the chemical solitariness of others, preclude the formation of groups having equal rank and precise group-characters: thus the proposed chemical classification, though convenient, has no pretension to scientific exactness. I suggest the following eight groups:

GROUP I.—THE OXIDES.

Zinc white	ZnO.	Cœruleum	CoO, n SnO_2.
Green oxide of chromium	Cr_2O_3.	Cobalt blue	CoO, n Al_2O_3.
Burnt umber	Oxides of Fe and Mn.	Venetian red	Chiefly Fe_2O_3.
Cobalt Green	CoO, n ZnO.	Light red	Chiefly Fe_2O_3.
		Indian red	Chiefly Fe_2O_3.
		Burnt Sienna	Chiefly Fe_2O_3.

Group II.—The Sulphides.

Cadmium yellow	CdS.
King's yellow	As_2S_3.
Vermilion	HgS.
Ultramarine; Artificial ultramarine	Contain Si, Al, Na, O, and sulphur.

Group III.—The Hydrates.

Yellow ochre	$Fe_2O_3nH_2O$, etc.
Raw sienna; Raw umber	Contain oxides and hydrates of Fe and Mn, etc.
Emerald oxide of chromium	Cr_2O_3, $2H_2O$.

Group IV.—The Carbonates.

Flake white	$2PbCO_3, PbH_2O_2$.
Whitening	$CaCO_3$.
Malachite	$CuCO_3$, CuH_2O_2.
Chessylite	$2CuCO_3, CuH_2O_2$

Group V.—The Silicates.

Terre Verte	Silicate of Fe, K, Mg.
Smalt	Silicate of Co, and K.

Group VI.—Other Inorganic Salts.

Baryta white	$BaSO_4$.
Aureolin	$K_6Co_2 12NO_2$.
Baryta yellow	$BaCrO_4$.
Strontia yellow	$SrCrO_4$.
Chrome yellow	$PbCrO_4$.
Chrome red	Pb_2CrO_5.
Zinc Chromate	$ZnCrO_4$.

Group VI.—Other Inorganic Salts (*continued*).

Naples yellow	Contains Pb, Sb, O.
Schweinfurt green	Contains Cu, As, O.

Group VII.—Organic Compounds.

Indian yellow.
Yellow lake.
Yellow madder.
Brown pink.
Rose madder.
Madder carmine.
Rubens' madder.
Madder red.
Purple madder.
Brown madder.
Carmine.
Crimson lake.
Scarlet lake.
Purple lake.
Indian lake.
Sap green.
Verdigris.
Emerald green.
Indigo.
Prussian blue.
Bitumen.
Bistre.
Sepia.
Vandyke brown.

Group VIII.—Elements.

Silver.
Gold.

Ivory Black; Charcoal black; Lamp black; Indian ink; Graphite	Contain carbon.

One of the chief lessons to be learnt from this classification is this, that the members of each class, as a general rule, exert no action upon one another. This is explained easily. The oxides of Group I., having already taken up the full complement of oxygen which they can acquire under ordinary conditions, are not likely to be oxidized by admixture with other oxides of similar character. In the same manner the sulphides of Group II. neither give sulphur to, or receive it from the other sulphides, for all of them have been produced in the presence of excess of sulphur. The following characteristics of each group may prove useful in the study of their chief members:

GROUP I.: THE OXIDES.—These have generally been prepared at a high temperature, and are not easily amenable to chemical or physical change: they are, moreover, not liable to affect other pigments, being practically inert.

GROUP II.: THE SULPHIDES.—Some of these may give up sulphur to the metallic bases of other pigments. Thus cadmium yellow blackens emerald green, producing copper sulphide. One of these pigments, vermilion, is prone to a molecular change, whereby the red crystalline form passes, without chemical alteration, into the black amorphous variety. The members of this group sometimes contain free sulphur, or injurious sulphur compounds.

GROUP III.: THE HYDRATES.—The water present in these compounds exists in two states, essential and hygroscopic. Sometimes a part of the former may be lost, and a change of hue occur in consequence, but the alteration is rare and slight. In the case of raw umber, the water present acts rather in aiding the oxygen of the air, under the influence of sunlight, to oxidize some of the peaty or bituminous matter sometimes present in this pigment.

GROUP IV.: THE CARBONATES.—Three out of the four

carbonates included in this group are liable to suffer change on account of the metal they contain (lead or copper) combining with sulphur, and so forming a brown or black sulphide.

GROUP V.: THE SILICATES.—These are generally inert bodies, little prone to suffer or cause change. Some of the ochreous earths contain silicates of iron, manganese, and alumina, as well as the hydrates of the two former metals, and so might be placed in this group.

GROUP VI.: OTHER INORGANIC SALTS.—A rather miscellaneous group. Some members, such as the chromates, with lead antimoniate, tend to lose a part of their oxygen, which may serve to attack some of the organic pigments of the next group with which they happen to be mingled; they themselves become greenish, or greyish at the same time. Those members which contain lead may become brown or black in the presence of sulphuretted hydrogen, etc. Aureolin, a nitrite of cobalt and potassium, owing to its nitrous constituent, acts upon indigo and some other organic pigments.

GROUP VII.: ORGANIC COMPOUNDS.—This group includes many more pigments than any other: not one of its members possesses the permanency belonging to the majority of the mineral pigments, while some are so fugitive that they may even be used for producing a photographic picture by being exposed to sunlight under a negative. This fading is generally due to the combined action of water and oxygen: in oily and resinous media it is lessened, retarded, or even prevented by the hydrofuge character of these vehicles.

GROUP VIII.: THE ELEMENTS.—All the black pigments in ordinary use consist of the element carbon more or less pure, and are not subject to change: graphite is a form of carbon absolutely unchangeable, and quite inert. Gold, if nearly free from alloy, is unalterable, but silver readily tarnishes by absorption of sulphur; drawings in silver-point are frequently found to have changed in hue through this cause.

It should be noted that members of each group, though presenting one or more characters in common, often exhibit certain chemical and physical differences of deportment. Here is a list of the chief changes which they are capable of suffering, with illustrative examples:

Molecular rearrangement ...	Vermilion.
Subsidence...	Smalt.
Volatilization	King's yellow.
Solution	Strontia yellow; aureolin.
Fusion	Asphalt.
Oxidation	Carmine; Vandyke brown.
Reduction	Naples yellow; chromates.
Sulphuration	White lead; emerald green.

The effect of pulverization upon pigments may be mentioned in this connection. Generally, the more finely an alterable pigment is ground, the more susceptible does it become to chemical injury: its colour becomes at the same time paler, and may even change in hue. Continued grinding, beyond the degree necessary to develop the proper colour, improves some pigments, but injures others.

No mention has been made in the present chapter or in the seven preceding chapters of certain beautiful pigments employed in this country or on the Continent. Experiments with such colours as Roman orange and madder green are now in progress, but it would be premature at present to pronounce any verdict as to their stability. It may, however, be at once stated that three different pigments sold under the name of green lake are quite untrustworthy.

CHAPTER XXI

TABLES OF PERMANENT, FUGITIVE, AND ALTERABLE PIGMENTS

By several different methods, data may be obtained which enable us to classify pigments—roughly, it is true—in accordance with their varying degrees of stability. Such data are derived partly from the known chemical and physical constitution of the various substances; partly from a study of old paintings and drawings in which they have been used; and partly from special experimental tests of permanency to which they have been subjected. Selections from these data are given in Chapters XX., XXIV., and XXVI. of the present work; but much additional information has been furnished by other trials, conducted by the author and other experimenters, for which space could not be found in this volume. Tables constructed from such data must not be regarded as affording exact values, but merely approximations. From some minute and often obscure cause differences of deportment, under exposure to hostile influences, will occasionally be observed in the case of two specimens of the same pigment having the same hue. And, further, the grouping of pigments into a small number of classes is a conventional and convenient arrangement which cannot accurately represent the numerous degrees of stability or instability which characterize the several pigments under discussion. For when we leave the practically unalterable mineral pigments, we have to deal with a number of preparations which fall by irregular and often barely recognisable steps from the almost permanent to the hopelessly fugitive. The action of mixed pigments upon one another, though not as frequent as it is supposed to be, creates another difficulty in our classification, so also does the medium employed in painting, which may either protect an alterable pigment from change or aid in its destruction. In fact, each method of painting, if really distinct, requires a special classification of the pigments to be employed in carrying it out.

In the annexed classification, a limit of three orders of stability has been adopted, the first class including the truly permanent pigments; the second class those which, though liable to a variable measure of change, may yet generally be allowed; and the third class those which should be definitely excluded from the palette:

CLASSIFIED TABLE OF PIGMENTS FOR OIL-PAINTING

Class I	Class II	Class III
	White	
Baryta white. Zinc white. Flake white.		
	Yellow	
Yellow ochre. Raw sienna. Naples yellow. Cadmium yellow. Cadmium orange.	Aureolin. Baryta yellow. Indian yellow. Strontia yellow. Chrome yellow.	King's yellow. Yellow madder. Brown pink; yellow lake. Gamboge. Zinc chromate.
	Red	
Vermilion. Indian red. Light red. Venetian red. Red ochre.	Madder carmine. Rubens madder. Rose madder. Madder red. Purple madder.	Crimson lake. Carmine and burnt carmine. Indian lake. Scarlet lake. Purple lake.

Class I	Class II	Class III
	Green	
Emerald oxide of chromium. Green oxide of chromium. Cobalt green.	Emerald green. Terre verte. Malachite.	Verdigris. Sap green. 'Green vermilion,' etc.
	Blue	
Ultramarine. Artificial ultramarine. Cobalt. Cœruleum.	Smalt. Prussian blue. Antwerp blue.	Indigo. Blue verditer. Blue ochre.
	Brown and Black	
Burnt sienna. Raw and burnt umber. Cappagh brown. Caledonian brown. Verona brown. Prussian brown. Ivory-black. Charcoal-black. Lamp-black. Graphite.	Madder brown. Cologne earth. Bitumen. Vandyke brown A. (*earthy*).	Vandyke brown B. (*bituminous*).

In order to adapt the foregoing classified table to water-colours, some changes and additions must be made. Flake white, Naples yellow (true), cadmium (pale), and vermilion (artificial), must be removed from the Class (I.) of permanent pigments and placed in Class III., to which also must be relegated several pigments from Class II., namely, chrome yellow, malachite, and madder brown. Of course, it should be clearly understood that no pigment belonging to Class III. should be employed in artistic painting. One satisfactory addition, and one only, can be made to Class I. in the table. Indian ink is a pigment available for water-colour painting, and when it is free from a brownish hue may be safely used. Bistre and sepia are likewise used only as water-colours, but they are both fugitive, and must be placed in Class III.

Almost the same modifications of the table are required in the case of tempera-painting as in water-colour painting. With fresco-painting the exclusion of many more pigments is an absolute necessity, as they are completely ruined by caustic lime. Not only are all the chromates inadmissible, as well as all the pigments which cannot be trusted as water-colours, but likewise Prussian blue and Antwerp blue, while the madder colours are much altered in hue when used in this process. In stereochromy the number of available pigments is still further reduced.

It may not be uninstructive if we cite in this place the classification of pigments as used in oil which M. Decaux has published. The order followed by this experimenter is that of stability; the figures prefixed to the names of the individual pigments indicate the degree of permanence, 1 marking out the materials which are quite unchangeable, while 45 is the most fugitive of all:

DECAUX'S TABLE OF PIGMENTS FOR OIL-PAINTING

CLASS I

1. Zinc white.
1. Flake white.
1. Yellow ochre.
1. Naples yellow.
1. Cadmium (deep).
1. Raw sienna.
1. Red ochre.
1. Mars red.
1. Venetian red.
1. Burnt Italian earth.
1. Green oxide of chromium.
1. Ivory black.
1. Terre verte.
1. Green ultramarine.
1. Cobalt blue.
1. Artif. ultramarine.
1. Ivory black.
2. Mars brown.
3. Burnt sienna.
4. Cobalt green.
5. Mars yellow.
6. Mars orange.
7. Burnt umber.
8. Viridian.
9. Indian red.
10. Mars violet.
11. Indian yellow.
12. Emerald green.

CLASS II

13. Malachite green.
14. Scheele's green.
15. Raw umber.
16. Vandyke brown.
17. Prussian blue.

18 to 23. Various madder lakes.

24. Madder carmine.
26. Madder 'rose doré.'
27. Brown madder.
29. Cassel earth.

CLASS III

30. Pale chrome.
31. Zinc chromate.
32. Pale cadmium.
33. Orange chrome.
35. Asphalt.
36. Brown pink.
38. Vermilion.
42. Burnt carmine.
43. Yellow lake.
44. Carmine.
45. Crimson lake.

On comparing this classified list with that previously given a general accordance will be perceived, the low positions given to raw umber and to vermilion, as well as the high places assigned to Indian yellow and to terre verte, constituting the chief exceptions.

CHAPTER XXII

SELECTED AND RESTRICTED PALETTES

It is by no means easy to construct a palette which shall be at once artistically and scientifically perfect. For it is impossible to exclude every pigment which is susceptible of change, and it is unwise to include every pigment for which the fancies and partialities of particular painters desire to find a place. An artist finds it easy to obtain a required hue by means of a special pigment, and is naturally reluctant to learn by tedious experimenting whether it cannot be secured by means of a more complex commingling of the ordinary paints. And although some great masters have done marvellous things with five, four, or even three pigments only, there is no sound argument which can be urged in favour of so severe a restriction. If much mixing of paints be bad, then a reasonable enlargement of the palette will render such mixing unnecessary. And the artist wants something more than a mere match in hue; he knows that there is a peculiar quality of colour to be sought as well. He can make a transparent pigment opaque, but the reverse operation is impracticable. Scumbling of one opaque colour thinly over another which is also opaque very imperfectly attains the effect of translucency. So the artist demands, in addition to a chromatic series of opaque pigments, a second series possessed of transparency, or, at least, of translucency.

Thus he adds to his cadmium yellow, aureolin ; to his vermilion, madder carmine ; to his emerald green, viridian ; to his cœruleum or cobalt, ultramarine. And, moreover, he has to take account of the peculiar and often unexpected effects produced by the lightening of the tone of a pigment by commixture with white, and by the darkening due to the addition of black. Two nearly identical translucent reds may yield with white two different hues, one verging on salmon, the other on rose. Charcoal-black yields with aureolin or Indian yellow a series of greens quite distinct from those obtained by mixing these yellow pigments with ivory-black. So the artist in making his first choice from the whole number of trustworthy pigments at his command, will proceed towards his final selection by two stages. He first retains those pigments which commend themselves to his judgment for their own chromatic qualities when unmixed; he then proceeds to test the characteristics of the remainder by trying the tints which they severally produce with white, the shades they yield with black, and the mixed hues to which they give rise by commixture with one another in twos and threes. To this set of experiments he adds another, in which these pigments are mixed, after the same manner, with those belonging to the first series. As the result of these trials the artist will be enabled to exclude several paints which would merely serve to encumber his palette.

Before deciding finally as to the elements which shall be retained for our fundamental palette, it will be instructive to study the selections of pigments which from time to time have been employed by artists of recent times and of the present day. The obvious weakness of many of such palettes lies in their inclusion of a few treacherous pigments, such as asphaltum, and of a few evanescent pigments, such as carmine, crimson lake, and the bituminous variety of Vandyke brown. Nevertheless, in making our selection of pigments from the classified list previously given, we may obtain many useful hints from the palettes employed by artists with whose works we are familiar. It is particularly interesting to observe how extremely restricted were the sets of pigments used by several painters who are distinguished for the refinement and for the rich variety of hues shown in their works. In the following paragraphs the names of all decidedly fugitive and alterable pigments are printed in italics.

Sir Joshua Reynolds, although too fond of varying his practice by the introduction of many dangerous compounds, and by the use, in the same picture, of incompatible media and methods, executed many works between the years 1770 and 1775 with one or other of these five restricted palettes, containing from four to eight pigments:

i. Flake white.	Yellow ochre.	*Lake.*	Ultramarine.	Black.
ii. Flake white.	Yellow ochre. *Orpiment.*	*Lake.* *Carmine.*	Ultramarine.	Black. Blue black.
iii. Flake white.	Yellow ochre. Naples yellow.	*Carmine.* Vermilion.	Ultramarine.	Black.
iv. Flake white.	Asphaltum.	Vermilion.	Blue.	
v. Flake white.	Naples yellow.	*Lake.* *Minium.*	Asphaltum.	

Paul Delaroche and H. Vernet employed these eleven pigments:

Flake white.	Yellow ochre. Naples yellow. Raw sienna.	Vermilion. *Lake.* Brun rouge. Burnt sienna.	Artificial ultramarine.	Blue black. Ivory black.

W. Etty, R.A., used ten pigments:

Flake white.	Naples yellow.	Vermilion. Light red. Indian red. *Lake.*	Terre verte.	Raw umber. Burnt umber. Black.

Samuel Palmer employed in oil painting the following pigments, eighteen to twenty in number:

Flake white.	Naples yellow.	Vermilion.	Terre verte.	Artificial ultramarine.
	Yellow ochre.	Venetian red.	Green oxide chromium.	Antwerp blue.
	Raw sienna.	Indian red.	Emerald green.	Cobalt.
	Aureolin.	Madder carmine, etc.		
	Cadmium yellow.			

He added to these, as water colours, *gamboge* and Prussian blue.

Thomas Wright, of Derby, employed fourteen pigments, and it is to be presumed flake-white also:

Naples yellow.	Vermilion.	*Carmine.*	Terraceum blue (?).	Ivory black.
Brown pink.	Burnt ochre.	*Lake.*	Ultramarine.	
	Indian red.	*Burnt lake.*	Prussian blue.	
	Light red.		Lake azure (?).	

From the *Portfolio* of 1875-6 we obtain the particulars given below concerning the pigments employed by some distinguished living artists. It should be added that in one or two cases these painters have since somewhat altered their selections of colours. The palettes quoted have been chosen as representative of different types.

Mr. A. employs ten pigments:

Flake white.	Naples yellow.	Vermilion.	Cobalt blue.	Burnt sienna.
	Yellow ochre.	Venetian red.	Antwerp blue.	Raw umber.
	Cadmium yellow.	Pink madder.		*Vandyke brown.*
	Raw sienna.			Ivory-black.

Mr. B. twenty-three pigments:

Zinc white.	Lemon yellow.	Vermilion.	Oxide of chromium.	Ultramarine.
	Cadmium, 1, 2, 3.	Light red.	Emerald oxide chromium.	Artificial ultramarine.
	Aureolin.	Indian red.	Cobalt green.	Cobalt blue.
	Yellow ochre.	Madder lake.		
	Roman ochre.			

With, in addition, raw umber, burnt sienna, cappagh brown, *Vandyke brown*, ivory black.

Mr. C. ten pigments:

White.	Yellow ochre.	Vermilion.	Cobalt blue.	Raw umber.
	Raw sienna.	Rose madder.		*Vandyke brown.*
		Purple lake.		Ivory black.

Mr. D. fifteen pigments in water-colour pictures:

Chinese white.	Yellow ochre.	Vermilion.	Cobalt.	Burnt sienna.
	Raw sienna.	Light red.	Artificial ultramarine.	*Vandyke brown.*
		Venetian red.	*Indigo.*	Ivory-black.
		Indian lake.	Prussian blue.	
			Antwerp blue.	

Mr. E. twenty-six pigments :

Flake white.	Lemon yellow. Cadmium yellow. *Gamboge.* *Laque Robert.* Naples yellow. Yellow ochre. Raw sienna.	Orange vermilion. Madder carmine. Venetian red. Indian red. *Indian lake.* *Burnt lake.*	Ultramarine. Antwerp blue. Cobalt blue. Green oxide chromium. Emerald oxide chromium. Malachite.	Raw umber. Burnt umber. Burnt sienna. Cologne earth. Blue black. Ivory-black.

Mr. E. has also used the two so-called *yellow madders.*

The selection of a good set of permanent or fairly permanent pigments must depend to some extent upon the idiosyncrasy of the artist, upon his training and methods of work, upon the class of subjects with which he deals. As a good general working set for oils, the following selection is offered. It is arranged in two sections, the second including what may be called 'supplementary' pigments :

Section I. includes 12 pigments.	Flake white. Cadmium yellow. Aureolin. Yellow ochre.	Vermilion. Madder carmine. Light red.	Viridian. Artificial ultramarine.	Raw umber. Cappagh brown. Ivory-black.
Section II. includes 13 pigments.	Raw sienna. Naples yellow. Baryta yellow.	Purple madder. Madder brown.	Green oxide chromium. Terre verte. Malachite. Emerald green.	Cobalt. Prussian blue, (insol.). Burnt sienna. Caledonian brown.

It will be remembered that emerald green cannot be safely associated with cadmium yellow, and that there is no reason why several other pigments should not be included in Section II., beyond the desirability of limiting the number of paints to those really required. Garance dorée, Rubens madder, cobalt green, burnt umber, Verona brown, vine black, and graphite might be added to the list. On the other hand, further restrictions become by practice possible. One does not know what white, vermilion, yellow, and vine or charcoal black can do until one has purposely debarred one's self from the employment of any other coloured pigments. Here are two such restricted palettes :

1. Flake-white, yellow ochre, light red, cobalt, ivory-black.

2. Flake-white, cadmium yellow, vermilion, ultramarine, ivory-black.

A third restricted palette, containing ten pigments instead of five, is thus constituted :

3. Flake-white, yellow ochre, cadmium yellow, aureolin, vermilion, madder carmine, ultramarine, viridian, Cappagh brown, ivory-black.

It is scarcely necessary to say that the capacity of No. 1 for representing the range of natural hues is extremely limited ; indeed, it is fitted only for 'dead colouring,' and for the 'first painting.' With No. 3, however, we can imitate with a near approach to exactness all the pigments excluded from this palette, and we may therefore regard it as practically complete. Some of the hues obtained by the mixtures necessary to employ for this purpose will be a little less luminous than the originals, since these hues will have been produced by the increased absorption of certain elements of the incident white light—they are consequently duller, or have more grey in them. This palette, No. 3, is nearly the same as one devised by Mr. P. G. Hamerton (*Portfolio*, 1876, p. 132), which was constituted of flake-white, pale cadmium, yellow ochre, vermilion, rose madder, artificial ultramarine, emerald oxide of chromium, Vandyke brown, black. I have added one pigment, aureolin, and have substituted for pale cadmium, full cadmium yellow ; for rose madder, the more stable madder carmine ; and for Vandyke brown, Cappagh brown. Mr. Hamerton tested the range of his restricted palette by imitating with its constituents many of the excluded pigments. I give some of his results, as modified by my own experiments with my palette No. 3.

Naples Yellow.—Imitated by flake-white, with cadmium yellow and a trace of yellow ochre : exact.

Lemon Yellow.—Flake-white, cadmium yellow, with a trace of viridian : less brilliant than the original.

Cadmium Orange.—Cadmium yellow, with vermilion : less brilliant.

Light Red.—Vermilion, yellow ochre, Cappagh brown.

Venetian Red.—Vermilion, yellow ochre, madder carmine, a little Cappagh brown : exact.

Indian Red.—Vermilion, trace of yellow ochre, madder carmine, ivory black : a good match, but less translucent.

Cobalt Blue.—Artificial ultramarine, flake-white, a little viridian : less translucent ; does not match cobalt blue by artificial light.

Prussian Blue.—Ultramarine, black, a trace of viridian : lacks the translucency and depth of the original.

Raw Sienna.—Yellow ochre, aureolin, Cappagh brown.

Burnt Sienna.—Madder carmine and Cappagh brown, with a trace of vermilion : less translucent.

Emerald Green.—White, cadmium yellow, viridian, artificial ultramarine : not so brilliant as the orginal.

Malachite.—White, cadmium yellow, yellow ochre, viridian, ultramarine.

Terre Verte.—White, aureolin, viridian, ivory-black.

Cobalt Green.—Ultramarine, viridian, trace of flake-white.

Indigo.—Ultramarine, with black and trace of viridian : very close.

Vandyke Brown.—Cappagh brown, with much madder carmine and a little ivory-black.

It is needless to multiply further our illustrations of the resources at the command of the painter who limits himself to our restricted palette of ten pigments (No. 3, page 243), as experimental trials of its capacity are easily made.

So far thus as regards selected and restricted palettes of oil-colours. Some modifications must be made in our list in order to devise corresponding palettes of useful and enduring water colours. In the more extended list (page 242), zinc-white must replace flake-white, while vermilion, purple madder, brown madder, malachite, and emerald green must be discarded. In the limited palette (No. 3), the changes to be made comprise the substitution of zinc-white (= Chinese white) for flake-white, the replacement of artificial vermilion by native cinnabar, and of Cappagh brown by Mars brown, and of ivory-black by Indian ink. The two palettes (A. and B.) will then finally assume the following forms:

(A.) Section I. includes 13 pigments.	Zinc white. Cadmium yellow. Aureolin. Yellow ochre.	Light red. Indian red. Madder carmine.	Viridian. Artificial ultramarine. Cobalt.	Raw umber. Burnt sienna. Indian ink.
Section II. includes 5 pigments.	Raw sienna.	Vermilion (cinnabar).	Prussian blue (insol.).	Mars brown. Ivory-black.

Doubtless artists will especially miss from this palette five pigments, namely, gamboge, rose madder, brown madder, Vandyke brown, and indigo. But after the overwhelming evidence adduced in Chapter XXVI. as to the want of permanence shown by these water-colour paints, one feels compelled to exclude them. Our second and more restricted palette (B.) is thus composed:

(B.) Chinese white.	Yellow ochre. Cadmium yellow. Aureolin.	Cinnabar. Madder carmine.	Ultramarine. Viridian.	Mars brown Indian ink.

Although it is obvious that with these limited palettes it is impossible to produce exact imitations of every excluded pigment, yet there are two considerations which must not be forgotten in estimating the influence of this defect on artistic painting. Foremost may be placed the fact that pigments are rarely employed wholly unmodified by admixture with others; then it must be noted that the differences between our imitations and the original pigments which they are intended to replace are rather those of lessened brightness, translucency, and depth than those of hue.

PART IV

METHODS AND RESULTS

Chapter XXIII.—Painting Methods. Chapter XXIV.—Study of Old Paintings and Drawings. Chapter XXV.—Conservation of Pictures. Chapter XXVI.—Trials of Pigments.

CHAPTER XXIII

PAINTING-METHODS

As the grounds, vehicles, and pigments employed in painting have been already described in Parts I., II., and III. of this volume, it will not be necessary to do more in the present chapter than give a summary or general view of the chemistry of each method of employing these materials.

These methods are six in number, and may be thus defined:

Methods	*Vehicles*	*Changes during Fixing*
1. TEMPERA - -	Egg-yolk emulsion; solution of gelatin or albumen - - - - - -	Desiccation or coagulation.
2. FRESCO - -	Lime-water, in both buon' fresco and fresco secco - - - - - -	Carbonation.
3. STEREOCHROMY	Aqueous solutions of alkaline silicates -	Formation of insoluble silicates.
4. OIL - PAINTING AND SPIRIT - FRESCO	Oil, and solutions of resin, wax, paraffin	Oxidation, Resinification, Evaporation, Solidification.
5. WATER-COLOUR	Aqueous solutions of gum, glycerin, honey - - - - - - -	Desiccation.
6. PASTEL, CHARCOAL, PLUMBAGO, SILVER-POINT	None - - - - - - - -	None.

1. *Tempera-painting*, or painting in distemper, is generally assumed to include two, if not three, methods of procedure, in which different vehicles or media are employed. These vehicles all contain a nitrogenous constituent; but in one of them—and that the most important—oil or fat is present in addition. Tempera-grounds must be rigid, tenacious, and

firm; they need not be dry, but if organic pigments are to be used, they should not contain caustic lime. Thus, a surface of plaster made with slaked lime and sand must have been so long exposed to the air as to have absorbed the amount of carbonic acid necessary to convert the hydrate of lime present into 'mild lime'—that is, the carbonate. To detect the existence of caustic lime in such a painting-ground recourse may be had to test-papers. Three kinds are available for this purpose. Thus, yellow *turmeric-paper*, first wetted and laid upon the surface of the plaster, should show no change of colour; if it become reddish, the presence of caustic lime is indicated. Under the same circumstances red *litmus-paper* turns blue or purple, while *phenolphthalein-paper* acquires a crimson hue. If these tests show the absence of caustic lime, the painting may be commenced, otherwise the surface must be carbonated by syringing it or washing it with water charged with carbonic acid gas. These precautions are, of course, unnecessary in cases where the painting-ground has been prepared with plaster of Paris or other neutral compositions of which caustic lime is not a component. Before commencing work the painting-ground must be slightly and uniformly moistened with distilled water, and then coated with weak size. The pigments to be employed are those recommended for use as water-colours; they are thoroughly mixed with the medium to be employed, namely, egg-yolk emulsion, or size, or prepared white of egg. These media serve not only to bind the pigments to the ground, but also the coloured particles to one another. To render the egg-yolk more tractable, its alkaline reaction should be exactly neutralized by the cautious addition of a very few drops of white vinegar—fig-tree sap or white wine were formerly employed for the same purpose. Some artists content themselves with diluting the egg-yolks with a little water; others add a small proportion of white of egg, previously shaken with a little water and filtered. To keep the egg-emulsion sweet, a lump of camphor or a few cloves may be put into it. Size and also white of egg have been employed in tempera-painting. The white of egg needs dilution with water, thorough shaking, and then filtering through muslin. When egg-yolk is used in this method of painting, the oil in it gradually hardens, while the albuminoid matters which accompany it become partly insoluble and coagulated. As the amount of oil in egg-yolk is twice as great (30 per cent.) as the albuminoid matters (15 per cent.), this vehicle presents considerable resemblance to those employed in oil-painting, the albuminoid matters corresponding in a measure to the resins often used in the latter method. This vehicle does not act so effectually as oil and varnish in 'locking up' pigments, and so the protection against change which it affords is less. Moreover, instances have been observed in which the sulphur present in the albuminoids of egg-yolk has acted injuriously upon some of the pigments of the picture; but by excluding, as we now do, all paints containing lead and copper from the tempera-palette, accidents of this kind are prevented. A finished tempera-picture was often—one might almost say generally—rubbed with a cloth, sized, and varnished, the varnish being often made by dissolving sandarac in oil. The tone of the colours was thus warmed, while further protection was at the same time afforded against moisture and impure air.

2. In *fresco-painting*—both buon' fresco and fresco secco—the ground must not only be wet, but caustic. In true fresco the pigments are applied to the last and freshly-spread coat of plaster before it has had time to absorb more than a trace of carbonic acid from the air; the painting-ground is in fact saturated with an aqueous solution of hydrate of

lime, while there remains a large reserve of this compound in an undissolved condition. When on such a surface a layer of pigment mixed with water is placed, as that water evaporates the lime-water in the ground diffuses into the paint, soaks it through and through, and gradually takes up carbonic acid from the air, thus producing carbonate of lime, which acts as the binding material in this method. As there still exists an ample reserve of hydrate of lime in the ground, wetting the painted surface with pure water will cause more of this hydrate to enter into solution, and so the liquid present in the plaster will be reinforced with a fresh supply of the binding material. Ultimately the ground and the pigment become incorporated and harden together. If more binding material be required, it may be introduced by means of lime-water itself, or even by baryta-water, which contains sixty times as much hydrate of baryta as the strongest lime-water contains of hydrate of lime; these liquids or hydrate of lime may also be mixed with the pigments used. Although the chief binding material in fresco-painting is this carbonate of lime, yet with some plasters and with some pigments another substance is produced. This is silicate of lime, produced by the action of caustic-lime in solution upon the soluble silica of the plaster or of the pigments. Some sands, infusorial earths, and ochreous pigments contain such soluble silica, but it is certainly not present in every case. Silicate of lime as a binding material is more permanent than the carbonate.

In fresco secco the plaster is allowed to harden, and, in some measure, to dry, and the operation of painting may be continued at leisure. The ground is moistened with lime- or baryta-water, and the pigments are mixed with one or other of these liquids, or with a little slaked lime. This modified process is far easier of execution than true fresco; but the fixation of the pigments, though resulting from the same cause, is less complete.

The protection afforded to the pigments by the binding material in fresco-painting is not generally very efficient. In the case of a dry wall, free from soluble saline matter, and exposed to a pure atmosphere, it may remain good for centuries. But in air contaminated with the products of the combustion of coal and gas, and with tarry and sooty impurities, a fresco picture soon perishes. The binding carbonate of lime is converted into the sulphate, breaking up the paint, and becoming itself disintegrated in the process of change. Through the same cause, and through the production of sulphate of magnesia from the carbonate of magnesia in the plaster, even the layer of paint itself may scale off, while the lodgment of dirt and soot upon the surface obscures such colours as still remain in their place. And fresco-paintings often show scaling-off, by reason of the interposition of a film of carbonate of lime between the coats of paint—a film formed during the completion of the picture.

True fresco did not come into use until the close of the fourteenth century. About the year 1390, Pietro d'Orvieto painted some subjects from Genesis in the Campo Santo at Pisa. In 1503, Pinturicchio, at Siena, began some works in fresco, which he finished in tempera with lakes and other pigments injured by lime. This mixed method was much used in Italy to a late period, for it enabled a greater richness of effect to be attained. For the palette of the painter in true fresco is severely restricted in certain directions, very few colours of organic origin withstanding the decomposing action of lime. It is a good plan to test each pigment intended to be employed in this method: The pure pigment is thinly painted over a slab of plaster of Paris, and then half of it moistened with lime- or baryta-water. No change of hue, only a lightening of the tone, should be observed, after drying, in the treated portion. Prussian blue may be named amongst the pigments most quickly and seriously altered by lime; it becomes a mere stain of rust. As lime in the caustic state acts strongly upon wood, it is necessary to employ palettes of zinc or glazed earthenware; bone or ivory palette-knives are preferable to those of steel.

3. In *stereochromy*, or water-glass painting, a process introduced more than forty years ago, the fixative employed is an alkaline silicate dissolved in water. From time to time different experimenters have improved the painting-grounds, the preparation of the pigments, and the mode of applying the fixing liquid; but the main chemical actions involved in this method of painting are identical in all the modifications which have been introduced. The constituents and preparation of painting-grounds adapted for this process have been discussed in Chapter II. The pigments should be treated, as recommended by Kuhlmann, with some of the fixing liquid, and then reground; in some cases they require the addition of oxide of zinc, powdered marble, powdered glass, carbonate of baryta, soluble silica, hydrate of alumina, etc., in order that their natural inaptitude for equal fixation by the alkaline silicate should be remedied. Opinions differ as to the desirability of treating the painting-ground with some of the water-glass solution before laying on the colours; but it is essential that if a solution of this silicate be used at this stage, it should be very dilute. The finished painting is sprayed with a warm dilute solution of potash water-glass or potash-soda water-glass, to which has been added liquor ammoniæ. The surface is shortly afterwards washed repeatedly with hot distilled water; and, if

necessary, the application of the water-glass solution, and the subsequent washing, are repeated. The final result of these operations is to bind the particles of pigment to one another, and to the ground, by means of an insoluble double silicate. This silicate, formed partly out of some of the constituents of the ground, of the pigments, and of the water-glass, mainly consists of silica, lime, and potash; it often contains zinc, magnesia, and alumina. The soluble salts removed by washing the painting with water are the carbonates of potash and ammonia; when, however, soda is present in the water-glass, carbonate of soda has been formed, and is removed at the same time. The pigments employed in stereochromy are more limited in number even than those available in fresco-painting, and consist chiefly of natural oxides and earths, the artificial oxides and hydrates of chromium and iron, cobalt green, ultramarine, cobalt blue, and ivory-black.

4. *Oil-Painting and Spirit-Fresco.*—The essential characteristic of these methods is to be found in the use of a binding material which is in itself insoluble in water. The painting-ground employed should be dry, and free from alkali and from soluble salts. If it be primed canvas or panel, it is a good plan to cleanse it with oxgall and water, or with a very weak solution of carbonate of ammonia, before commencing work. A discoloured lead-priming should be restored to its original brightness by laying a sheet of white blotting-paper upon it, and then just saturating this paper with a solution of peroxide of hydrogen. When the paper has become dry, it may be removed, and the bleaching of the tarnished ground will be found to have been effected, the brown sulphide of lead having been oxidized into the white sulphate. In order to learn whether a plaster-ground or a wall is sufficiently dry to be safely

painted upon in oil or spirit-fresco, the gelatin-test may be employed. A small oblong piece of coloured sheet-gelatin is held firmly and closely against the plaster or wall, by means of a stick applied at the centre. If hygroscopic equilibrium have been established between the wall and the air, the gelatin will remain flat; if the wall be moister than the air, the sheet will curl outwards, the inner surface becoming highly convex. Slate and several other suitable painting-grounds may be dried and further prepared for work in oil or spirit-fresco by heating them gradually in a water-oven up to the temperature of boiling water, and then rubbing them with a piece of hard paraffin-wax.* The slate is again heated in the water-oven, withdrawn, and then at once rubbed with a dry, warm cloth, so as to remove all excess of paraffin-wax. Other methods of treating stone, etc., for the reception of oil-colours have been previously given. A very convenient means of neutralizing the residual alkalinity of a lime-plaster ground intended for oil or spirit-fresco painting is afforded by linoleic acid.† This liquid fatty acid is an article of commerce, moderate in price, and easily obtainable. A wide-mouth tin of it is placed in a vessel of boiling water; when the linoleic acid is hot, it is paid on to the surface of the plaster with a wide brush, any excess being removed by wiping the ground with a cloth. Solid stearic acid may be melted and used in the same way, but its effect is inferior.

The vehicles employed in this method of painting are not miscible with water—are, in fact, *hydrofuge* materials repel-

* The hardest kind of paraffin is obtained from ozokerite. It is known in commerce as ceresin. Some samples, the best for the present purpose, require a temperature of over 80° C. to melt them.

† By linoleic acid is here meant the mixture of fatty acids obtainable from raw linseed oil.

lent of moisture. If an absorbent ground or other porous material be soaked with water, and then covered with oil, as the water evaporates the oil penetrates, and at last completely takes its place. But, on the other hand, the reverse process cannot be carried out, since the water outside will not displace the oil inside. These vehicles are either oils or else solid substances in solution—solids which, though insoluble in water, may be dissolved with more or less ease in one or other of a long series of liquid solvents (Chapters V., VI., XI. and XII.). The changes experienced by these vehicles and their constituents during the painting process may be thus summarized:

(*a*) The oils used absorb oxygen from the air, increasing in weight thereby to the extent of 8 or 9 per cent.—such increase in weight being accompanied by a considerable increase in bulk. This latter change is clearly shown when a layer of a drying oil, spread upon glass, is allowed to dry; it then becomes rippled or wrinkled from expansion; such expansion, owing to the viscosity of the oil, takes place mainly in a direction at right angles to that of the surface of the glass.

(*b*) The resins present in varnishes and media contract for some time after the major part of their volatile solvent has escaped by evaporation, and thus leave a residue which becomes fissured. In a properly-proportioned medium this contraction should be balanced, or rather more than balanced, by the expansion of the oil present. Hence the desirability of associating a varnish (or a resin dissolved in a volatile solvent) with a drying oil, in this method of painting.

(*c*) Waxes and solid paraffins, when once deposited from a solution by the escape of the solvent, neither expand nor contract by desiccation or oxidation, but only through changes of temperature.

(*d*) Most of the liquid solvents simply evaporate, leaving no fixed residue due to their previous presence. But oil of turpentine (spirits or essence of turpentine) generally behaves differently. Some kinds of oil of turpentine differ from the majority in this particular, but the remainder suffer two simultaneous changes. A portion evaporates; another portion absorbs oxygen from the air, becoming converted into a sticky, yellow, and resinous substance, which remains behind. The resin thus formed is a very objectionable constituent in the structure of a picture, and its production should be avoided either by employing a variety of turpentine not subject to easy resinification, or by using a freshly-distilled turpentine which has been secluded from the air, and in which a few lumps of freshly-burnt lime have been placed, to remove water and such resinous matters as may be produced.

An important precaution to be observed in the 'conduct' of a painting during its progress is based upon the two actions just referred to, namely, the oxidation of the oil during its hardening, and the escape of volatile solvents. The latter action takes place more easily than the former, and so if a picture is to be carried on rapidly to completion, the earlier and lower paintings should contain *less* oil than those nearer the surface, into which *more* oil and less resin (copal or amber), dissolved in some volatile solvent, should be introduced. If the reverse order be followed, the highly oleaginous layers below, having had no sufficient opportunity for oxidizing, drying, and hardening, will be rent by the strong and quickly-drying resinous layers above them.

The harder resins, paraffin-wax, wax, and oil, possess, in varying degrees, the power of 'locking up' the pigments with which they are mingled, in such a way that these become much less liable to act upon one another, and to suffer injury from external agencies. In a measure they repel and exclude moisture and oxygen—two of the chief agents of chemical change. But the value of these 'locking-up' materials has been exaggerated: they often prove quite ineffectual in preventing the oxidation or other change suffered by non-permanent pigments and the inter-action of pigments. For instance, the oil which surrounds each particle of cadmium yellow and emerald green, in a mixture of these two oil paints, is not capable of preventing the formation of the black sulphide of copper. And Mr. A. P. Laurie has recently found that when a layer of linseed oil is interposed between these oil-colours separately spread, it is the emerald green which appears to travel towards the cadmium yellow—perhaps owing to its solubility in the medium. In consequence, the production of spots of black sulphide of copper occurs chiefly, if not entirely, on that side of the oil-layer which is in contact with the cadmium yellow. To Mr. Laurie we are also indebted for a very ingenious method of comparing the locking-up function of various oils and resins. Mr. Laurie prepared some anhydrous sulphate of copper which is white, but acquires a blue colour when exposed to moisture. He ground this white sulphate with various media, painted glass slides with the mixtures, dried them in a desiccator, and then exposed them to moist air. A solution of amber in turpentine proved superior in its power of resisting the access of moisture to boiled linseed oil, oil-copal varnish, amber dissolved in oil, resin or mastic dissolved in turpentine. Another set of trials, in which the test substance was ground in linseed oil, allowed to harden in a desiccator and then coated with different varnishes, indicated a temporary superiority on the part of mastic in turpentine, and of oil-copal varnish over amber or copal in turpentine. The inferiority of the latter solutions may be due to the rupture in continuity of the resinous films which they leave on evaporation. (See *Journal of Chemical Industry*, June, 1890.) But it must not be forgotten that many an old oil-picture furnishes distinct evidence of the value of resinous matters (such as Venice turpentine) in locking up such changeable and destructive pigments as verdigris and orpiment. The slow and laborious execution of such paintings constituted an important element in the success achieved, for each layer dried and hardened before the next was applied.

It should be noted that different oil-paints contain very different percentages of oil, ranging from about 12 per cent. in flake-white to 67 per cent. in aureolin as commonly prepared. This fact should be taken into account, as far as possible, in adjusting the amount of resinous matter to be introduced during the course of work upon an oil-picture. A table giving approximately the percentages of oil present in a number of the oil-paints in common use will be found on page 47. Further information concerning such pigments is given in Chapters XIII. to XIX.

In completing an oil-picture, the three operations of 'glazing,' 'oiling out,' and 'varnishing' remain to be considered. As to glazing and oiling out, it should be stated that drying oil, with a *little* copal or amber varnish, should alone be employed—mastic varnish should never be added to the oil. Of course oil-paints are used in admixture with oil and copal for glazing purposes. If mastic be introduced, a risk is incurred of its partial removal during any cleaning operation to which the picture may be afterwards subjected. The question of the kind of varnish to be finally applied to an oil-picture has been much discussed. Our choice lies between a strong irremovable varnish, and a weak one capable of being abraded by friction, or of being dissolved by the application of a suitable solvent, which will not touch the true painting beneath. Mastic dissolved in turpentine fulfils the latter conditions; copal or amber dissolved in oil and thinned with turpentine, and mixed with a little oil, constitutes a strong, hard, irremovable protection to the surface, and becomes a part of the picture itself. Under no circumstances should any varnish be applied to the painting until the latter has become thoroughly hard and dry; the danger of tearing the layers of paint by such application will then have been reduced to a minimum. A further advantage of delay in varnishing a picture accrues through the increasing insolubility with age of the oxidized oil present therein, the pigments associated therewith becoming less liable to removal by any treatment to which the work may afterwards be submitted.

The chemistry of Gambier-Parry's spirit-fresco method, and of the process in which paraffin-wax and copal varnish are employed as the vehicle, is essentially the same as that of oil painting. The wax or paraffin-wax is introduced merely to secure a matt surface. Pictures executed in these methods are, of course, never varnished. The method of spirit-fresco was devised by the late Mr. Gambier-Parry with the object of obtaining such effects in mural paintings as are realized in true fresco, but with greater ease in working, and greater permanence under adverse atmospheric conditions. He desired to exclude linseed or other fixed drying oils completely from the medium and other materials employed. With this end in view, he directed that the pigments used should be ground up, not with oil, but with the medium itself. He was apparently unaware that the copal varnish, which enters largely into the composition of his vehicle, contains a greater proportion of oil than of any other ingredient (see Chapter XII.). So, after all, the medium used in spirit-fresco differs from that generally employed in oil-painting rather in the proportions than in the nature of its ingredients, so that in working with it we shall find that its binding character is obtained as a result of the same two changes which cause the fixing and solidification of an oil painting, namely, the oxidation of the oil, and the desiccation of the resin. The wax present suffers no chemical alteration, but merely solidifies. It should be added that the painting-ground for this method of working is first prepared with the medium diluted with oil of turpentine (see Chapter II.).

The method of painting with the paraffin-copal medium involves the same chemical and physical changes as those which occur in the use of the spirit-fresco vehicle, and is carried out in the same manner. Colours ground in oil may be used if any excess of oil present be first removed: dry colours ground in the medium are, however, to be preferred. The medium may be diluted to any desired consistency with spirits of turpentine or with oil of spike, but no dilution further than that required to secure perfect freedom in the manipulation and use of the paints is desirable.

5. *Water-colour Painting.*—The usual binding material in this method is gum; glycerin and honey are also employed to some extent. Raw honey should never be used, but only one of the sugars it contains, known to chemists as laevulose (Chapter VIII.). Great care must be taken not to introduce any unnecessary excess of either glycerin or laevulose, as these materials attract moisture from the air, and we know that moisture is one of the most potent agents in causing injury to works in water-colour. Glycerin and laevulose are, however, useful, when employed in moderation, for preserv-

ing the pigments in working condition, and in counteracting the tendency of gum to crack. The media used in water-colour painting, consisting wholly of aqueous solutions, afford very slight protection to the pigments used. In the presence of the moisture of the ground (paper often contains naturally 10 per cent. of water) and of the air, water-colour pigments have abundant opportunities, not only of acting upon one another, wherever from their chemical constitution such action is possible, but also of being acted upon by external agents. Thus it comes to pass that several pigments (artificial vermilion, for instance, and emerald green) useful in oil-painting cannot be safely used as water-colours. Again, there are a few pigments (such as strontia yellow) which are soluble in water, and which consequently may gradually sink into the paper, and so partially disappear from the surface.

Assuming the paper-ground to be of linen pulp, and free from 'filling,' from bleaching substances, from antichlors, and from fragments of iron, it will still contain about 5 per cent. of size. When in preparation for painting it is moistened with water, this size swells, and on the subsequent application of washes of pigments, enters partially into mechanical union with them, so that the various coloured materials applied to the surface become associated with the size rather than with the paper-fibres. One paint, Indian ink, itself contains size, and for this reason when washes of it are laid upon paper previously damped, their incorporation with the size of the latter is so intimate that their removal is impracticable. The size in a water-colour drawing becomes in time partly coagulated and insoluble; the gum merely dries. Instances are known where the size has in some degree ultimately perished.

6. *Pastel*, *Charcoal*, *Plumbago*, *Silver-point*.—The com-

mon characteristic of all the processes which form our sixth group is the absence of any vehicle or binding material. The usual ground on which drawings in the above-named substances are executed is paper (Chapter I.); but as the hold of coloured chalks and of charcoal is very precarious, the paper is generally mounted on some comparatively rigid backing, such as millboard, cardboard, copper, or panel. If a chalk or charcoal drawing be carried out on paper which has first received a wash of gum-water or of dextrin-solution, it is easy to effect a partial fixation of the powdery pigment by subsequently steaming the finished work, although it is usual to employ a fixing solution in the form of very fine spray to the finished drawing. For pastel work a specially prepared paper is now generally employed. This has a surface of finely-powdered pumice, which affords an efficient *tooth*, and helps in securing the coloured chalks. This result is further aided by the plan of working in and mingling the pigments by means of rubbing with the fingers and the palm of the artist's hand. Pastel-paper is often made of inferior pulp, and lacks strength. It should be less sized than paper intended for water-colours. Pastel colours are generally made with a basis of purified chalk mingled with the usual pigments in powder, a slight degree of cohesion being secured by making up the crayons with starch-paste or gum-tragacanth. One method of fixing the coloured particles in the finished drawing is to give the *back* of the paper a coat of isinglass solution. This solution is made with 1½ ounce of the best Russian isinglass dissolved in 1 quart of boiling distilled water. The liquid is filtered hot; when it is cold its own volume of proof spirit is added. In plumbago (lead-pencil) and silver-point work the mechanical adhesion of the coloured particles is naturally less imperfect, a portion of the plumbago or silver becoming, in fact, incorporated with the fibres of the paper-ground. This is particularly the case with silver-point, in which method the ground receives a particular preliminary preparation. One of the best materials for this purpose is Chinese white (oxide of zinc). An even wash of this pigment in the form of 'moist' water-colour is first spread over the paper. As a silver-point drawing is often heightened with touches of Chinese white, it is desirable to bring these into prominence by tinting the ground. For this purpose a small quantity of some permanent pigment is mixed with the wash of Chinese white. Yellow ochre, raw umber, green oxide of chromium, Mars violet, ultramarine with a little ivory-black, may be used. The 'tooth' of surface which increases the attrition of the silver-point is, however, furnished by the presence of the Chinese white. It should be added that the silver used should be free from any alloy of copper, which hardens the metal, but may advantageously contain a few per cents. of metallic lead; an alloy of 2 parts of lead with 1 part of tin was sometimes used instead of silver. The silver in silver-point drawings is liable to become brown from the sulphur compounds in impure air. The blackening of the high lights in old silver-point drawings is due to the tarnishing of the lead white employed; it may be got rid of by keeping the drawing for some time in an atmosphere of moist ozone.

Pastel or coloured chalk drawings frequently show a higher degree of preservation, so far as certain hues are concerned, than contemporary works executed in oil. One can easily account for the pure and fresh air of old pastel drawings, knowing that they have been carefully mounted and framed, and that there has been no oil or resin to yellow and darken the pigments. But how can the remarkable state of preservation in which the 'carnations' are found in so many examples be explained? Has the intimate commixture of chalk with crimson lake preserved the latter from the destructive action brought about by light? If there have been such a preservative action, has it been physical rather than chemical? Answers to such questions must be reserved until the chemistry of coloured chalks has been thoroughly studied.

CHAPTER XXIV

THE STUDY OF OLD PAINTINGS AND DRAWINGS

THE study of old pictures, with the view of discovering the causes of the physical and chemical changes which have taken place in them, is fraught with interest. The material on which they are executed, the medium employed, the pigments which can be identified, and the varnish which has been applied to the surface, all these matters demand attention. The dates of the various works examined, the countries in which they have been produced, the conditions under which they have been preserved, and the treatment to which they have been subjected, constitute elements in the investigation which, whenever possible, should be kept in view. But the adequate treatment of this extensive subject requires not a brief chapter, but a whole volume. And then our materials, though in some directions most abundant, are in great measure inaccessible. We must confine our attention to such specimens as are shown in our public galleries. Even then we find ourselves hampered by the impossibility of making the thorough investigation which is desirable, and by the too frequent absence of certain important data. In the present chapter we limit ourselves to some general remarks, and to a few brief observations upon a certain number of pictures in the National Gallery, the National Portrait Gallery, and the South Kensington Museum;

and our selection will be confined to paintings in oil, tempera, and water-colour, as the available works in fresco in England are too few and too fragmentary to furnish the information for which we are seeking.

It will hardly be necessary, with respect to changes in painting-grounds, to do more than refer the reader to what has been already said on this subject in Part I. of the present work. The causes of the decay of panels and of the convexity which their painted surface shows in so many cases have been already discussed. The disruption of the ground and of the superposed layer of paint consequent upon this convexity need no further explanation. The staining of the white priming which has been laid on certain kinds of wood has been traced to dark-coloured exudations of soluble organic matters. The grain of some kinds of wood, notably of oak in pictures of the Dutch and Flemish schools, often becomes painfully conspicuous in course of time, and gives to the surface-cracks a peculiar character. The microscopic structure of certain woods and the peculiar distribution of their histological constituents serve to explain these appearances. The causes of the decay and cracking of gesso-grounds in which size has been used, and the injurious mechanical and chemical alterations which paper and primed canvas may exhibit, have been already touched upon. The other conspicuous changes which may be observed in old pictures are connected with the medium, the pigments, or the varnish. All these matters have been referred to in Parts II. and III., yet there are three points on which further discussion may not be out of place. I refer to the number and character of the pigments used in early works, to the manipulation of the paint, and to the employment of white lead. Now, the pigments to which the earlier painters were restricted were not only few in number, but were mainly of mineral origin. At the first glance one sees that the Italian artists of the thirteenth century, and of the first half of the fourteenth, worked almost exclusively in natural inorganic pigments, two of which stand out in their works in startling prominence, namely, vermilion and ultramarine; and their pigments were nearly all opaque or semi-opaque. The absence of any pure and brilliant yellow, opaque or transparent, from their pictures is another noticeable characteristic. In the works of Jan van Eyck and Rogier van der Weyden, and in those of many of the Italian painters of the fifteenth century, the range of colours is more extensive. Pigments which could not be used in tempera or size, or which were semi-opaque when employed with these vehicles, gave great richness and variety to their works in oil. This tendency to press into the service of pictorial art other coloured materials besides those of mineral origin, namely, animal and vegetable pigments in considerable variety, became more marked as time went on. And during the last century and the present the progress of synthetical chemistry has placed at the disposal of the picture-maker a long series of pigments—good, bad, and indifferent,—so that the chances of introducing dangerous and fugitive colours have been enormously increased. It is to this increase in the number of pigments, and to their greatly extended range of composition, rather than to their mode of preparation, that one should attribute in great part the frequent deterioration of modern paintings.

But the second point to which reference has been made is concerned with the mode of laying on colours. The exquisitely minute and careful manipulation of Jan van Eyck, of Fra Giovanni Angelico, of Hans Memlinc, of Gerard Dou, of Gerard Terborch, and of many another old master, could not have been hurried. It was solid but smooth, and hardened gradually into one organic whole. And we could name several oil-painters of the eighteenth century, and even of the present day, whose work is executed in the same safe manner, and which, were it not for the occasional introduction of dubious materials, would be sure to remain sound for hundreds of years, provided, of course, that the painting-ground be satisfactory. But this careful mode of painting does not suit the temperament, nor is it capable of expressing the ideas of many artists. The thick impasto and loaded colour, the effective brush-work, the juicy pencil, and the dashing haste of several painters often prove to be elements of danger.

The third point, concerning which a few remarks seem advisable, is connected with the use of flake-white. There are many old oil-paintings in which the only perfectly-preserved parts of the work are those in which flake-white has been used with considerable freedom. Here the continuity of the layer of pigment is intact, elsewhere there are cracks and roughnesses and scalings-off. To what cause is the preservation of the high lights and of the paler flesh-tints attributable? The association of hardness and cohesiveness which these parts show is traceable to the white lead. This pigment was formerly always prepared in such a way as to contain a considerable quantity of lead hydrate. The particles of this hydrate do not lie, as it were, side by side with those of the chief constituent (the lead carbonate), but are so united with the latter as to form one complex compound. This compound acts upon the linseed or other drying oil with which it is ground, forming a lead-soap.

This substance seems to be formed in part out of the free fatty acids of the linseed oil, in part out of the glycerides present therein. It combines in an unusual degree toughness and hardness; and it is to the production of this lead-soap that I attribute the preservation of the continuity of surface of the whites, and of the pale tints into which white lead enters, in many an old picture. No other pigment in common use is capable of 'saponifying' the admixed oil to anything like the extent that characterizes white lead. Now there are modern preparations of white lead made chiefly by precipitation or the 'wet way' which produces a pigment containing little or no lead hydrate. Some writers on pigments advocate the use of these newer pigments. 'Why,' say they, 'should you carefully exclude from your pictures oils, and varnishes, and siccatives which contain lead in the form of a lead-soap, and then introduce the very same substance in your white lead ground in oil?' Many years ago I tried to answer such a question as this by means of experiment. I was actuated by a desire, based on theoretical considerations, of preventing altogether the formation of lead-soaps. I tried comparative experiments with zinc oxide, pure lead carbonate, and the Dutch-made lead hydrato-carbonate, or ordinary flake-white. The two lead pigments (with which alone we are now concerned) were washed thoroughly with distilled water and dried before being ground in linseed oil. The oil-paints thus prepared were spread in duplicate series upon glass, paper, and primed canvas; one set was kept in a dark box, the other was exposed to strong light. So decided was the superiority of the ordinary flake-white over the pure carbonate, when both series of specimens were examined after the lapse of various intervals of time, that I was reluctantly compelled to abandon my recommendation of the latter. Ease in working, solidity of body, and rapidity of drying, were not the only points of superiority; for the films of paint, after having been kept a year, showed differences in hardness and in smoothness of surface which were all in favour of the hydrated carbonate.

No discolouration was observed in the specimens exposed to light, except in the case of the pair upon paper; the absorbent ground had withdrawn some of the protecting oil, and both specimens had equally darkened. In darkness all the specimens had become of a somewhat greyish yellow, the discolouration being about equal in all the pairs, the pair spread on paper having, as in the previous case, become darker than the others. The late Mr. G. W. Wigner tried a somewhat similar series of experiments, and came to the same conclusions. I should add that these deductions were corroborated by the results of other trials, in which numerous permanent coloured pigments mixed in pale tints with these two lead whites were treated in the same way. If, however, we feel bound to recommend the ordinary flake-white instead of pure lead carbonate, that recommendation does not prevent us from excluding lead-containing oils from our pictures, seeing that we possess perfect substitutes for them, and that there is no reason for thus multiplying the causes of possible change.

Before commenting on some of the lessons to be drawn from individual pictures, it may be desirable to make a few observations on some of the changes frequently observable in old illuminated manuscripts and choral books. The tarnishing of lead and copper pigments laid on without any protection but that of gum is very frequently seen. The darkening of vermilion is apparently capricious,* but this molecular change has been explained when describing that pigment. Ultramarine always stands out absolutely intact; sometimes it acquires extraordinary prominence by reason of every other pigment on a page having altered. The red cochineal and kermes lakes have either gone or become

* An initial letter in vermilion, painted in the fifteenth century, and perfectly unchanged, became black by one year's exposure to sunshine.

paler and brownish. Sometimes fruits painted in vermilion have been shaded or dotted with a crimson lake, but the latter has disappeared, leaving nothing but a slight gummy appearance upon the scarlet ground. Blue flowers painted in smalt and veined with indigo show nothing but traces of the latter pigment. Verdigris, which is partly soluble in water, has run and discoloured the vellum, and at the same time has acquired a brownish hue. Sap-green, from buckthorn berries, has faded greatly. Lilien-grün (of the seventeenth century), from the flowers of *Iris germanica*, has disappeared.

We now cite a few pictures, out of a large number which have been studied for the purpose of observing the present state of the materials which have been used in their production. We begin with some works in the National Gallery.

Rogier van der Weyden (1400?-1464). No. 664. Painted on linen, which has received a *very thin* priming; the preservation of the work, although it has never been varnished, is practically perfect. From certain peculiarities in the touch, and from the minutely-wrought details of the woody landscape, I conclude that the medium used could scarcely have been the usual tempera containing egg-yolk, but was rather a thin size.

Melozzo da Forli (1438-1494). No. 755. It is probable that verdigris was employed in painting the green carpet in this work. If so, the cracks in this part of the picture (more conspicuous here than in other parts) would be due to the corrosive action of this dangerous pigment.

Jan van Eyck (1390?-1440). No. 186. The green robe in this famous oil picture shows distinct cracks, which differ in kind and are larger in size than any others in the work. The flesh tints are perfectly preserved as to texture. I suspect verdigris has formed a constituent of the green paint employed.

Tuscan School. No. 781. The lining of the cloak of Tobias in this picture seems to have been painted with verdigris; it is now very dark, in parts nearly black, although the pigment used has evidently been mingled with much protective resin, as its thickness is excessive when compared with that of other parts of the work.

Fra Giovanni Angelico (1387-1455). No. 663. The translucent reds and purples in this work have faded somewhat; the green, which appears to be malachite, has stood. On the whole this exquisite work in tempera is remarkably well-preserved.

Bennozzo Gozzoli (1420-1498). No. 591. The vermilion in this tempera picture is preserved in startling brilliancy; the translucent reds have become rather faded and embrowned.

Ridolfo del Ghirlandaio (1483-1561). No. 1,142. This oil picture, originally painted on wood, but now transferred to canvas, has been repaired and repainted in places. But the red glazings, apparently madder-lake, and the green colour, seemingly verdigris on malachite, are, if original, well preserved. The preservation of verdigris when glazed on malachite is not unusual; the two pigments are closely related chemically to one another, and are not likely to react.

Michelangelo Buonarroti (1475-1564). No. 790. This unfinished tempera picture illustrates the permanence of terre verte and madder-lake.

Gheeraert David (1450?-1523). No. 1,045. The fine crimson glazings and madder-lake in this oil picture are in perfect preservation.

Gregorio Schiavone (fl. 1470). No. 630. The madder

and vermilion in the robe of one of the figures in this tempera picture are well preserved. This is also the case in another work by the same artist, now in the author's possession.

In the National Gallery there are eleven portraits in wax-pigments from the Hawara Cemetery in the Fayum, Egypt. Seven of these seem to show the Tyrian or Murex purple, and prove that it has lasted apparently unchanged for some seventeen centuries. But it must be remembered that these remarkable paintings (Nos. 1,260 to 1,270) have been preserved in darkness since they were executed by Greek artists in the second century, or in the first half of the third. It has been conjectured that a pink colour occurring in some of these pictures was derived from madder. The other pigments were yellow, and red and brown ochre, charcoal black, a blue made from cupreous glass, and, possibly, red lead. The pigments, incorporated with wax, were laid on, in a fused condition, upon a distemper priming.

As to British pictures in the National Gallery, we can afford space for a few words only. The works of Sir Joshua Reynolds generally show the fading of the crimson lake (from cochineal) in the flesh tints, the vermilion and mineral yellows alone remaining. The picture of the 'Infant Samuel' may be cited as an example of the large and wide cracks caused by the free use of bitumen (in the dark background). Two paintings by J. M. W. Turner may be particularly mentioned. In No. 534, 'Chichester,' the bright lake-reds in the sky have become reduced to brown stains—anything but luminous. In No. 560, the 'Approach to Venice,' we notice how, in a group of small clouds near the top of the picture, where vermilion and lake have been introduced, the vermilion remains, but the lake is now a pale yellowish brown.

We may now pass on to some instructive examples preserved in the National Portrait Gallery:

Marc Gheeraedts (1561-1635). No. 64. In this portrait, painted probably in 1614, while the vermilion has stood, the translucent reds appear to have faded and changed. The white sleeves of the dress are ornamented with small sprigs, which are now brown, and were probably originally painted in some vegetable yellow. The reddish pattern of conventional foliage on the cloak now clashes with the colour of the chair, the curtain, and the table: the hues of all or some of these parts must have altered.

Sir Peter Lely (1617-1680). No. 509. A well-preserved picture in most respects, but it is probable that crimson-lake has been used for the satin dress, which is now a pinkish grey, and clashes with the flesh tints. This work was probably painted about 1669.

William Hogarth (1697-1764). No. 289. This portrait of Hogarth, painted in 1758 by himself, affords some information as to the pigments he employed. He holds in his left hand a mahogany palette 'set' with eight colours. The first of these is white lead, and remains unchanged; so also is the second, vermilion; the third is a pale warm brown, precisely the hue of faded crimson lake; the fourth is now nearly black and undeterminable; the fifth is yellow ochre, slightly embrowned; the sixth is a pale yellow, well preserved—much like true Naples yellow; the seventh is a grey-blue, probably much changed; and the eighth and last a fair lavender blue colour. The seventh pigment may have been indigo, and the last blue verditer. The cap on the artist's head has certainly faded in colour; probably it was painted with lake.

William Hoare, R.A. (1706-1792). No. 112. This portrait of Alexander Pope, in coloured crayons on grey paper, shows the blues apparently intact.

Thomas Phillips, R.A. (1770-1845). No. 269. This portrait of Faraday, painted in 1842, shows a large number of cracks, many of them wide. Where flake-white has been introduced somewhat freely, as in the face and hands, the shirt and collar, and the galvanic battery on the table, the paint has not lost its continuity.

John Partridge (1789-1872). No. 342. 'Meeting of the Royal Fine Arts Commission.' This picture was painted in 1846. Almost every part of it is very badly cracked through the use of bitumen, and perhaps also of much megilp. Even the shaded portions of the faces have not escaped, although the high lights have been preserved where the proportion of white lead present has been large.

The remaining works to which attention is now called are a few of the water-colour drawings shown in the South Kensington Museum.

William Daniell, R.A. (1769-1837). In one of the water-colours by this artist the blue and green elements have disappeared, save just in one little bit of smalt blue in a sailor's clothes. In another drawing the sky is now a mere dirty laboured stain.

William Green (1761-1823). No. 685. An iron red has here become too prominent.

Samuel Howitt (1765-1822). No. 3,019, 1876. Apparently well preserved, but, on further scrutiny, it seems that the ultramarine in the shaded parts of some of the rocks and trees stands out more prominently than it could have done originally.

Anne Frances Byrne (1775-1837). No. 1,358, 1874. In this fruit and flower piece the indigo in the sky has gone, while the yellow pigments and the red lakes have suffered greatly; the roses are blanched, and the purple grapes have lost their crimson element.

John James Chalon, R.A. (1778-1854). No. 570. The 'River Scene in Devonshire,' painted in 1808, shows the shaded parts of the clouds pink, from the loss of indigo. The hills are now too pale for the trees dotted upon them, through the fading of sap-green and gamboge, which have been used in painting the grass. The shadows under the ripples of the water have greatly altered, from the change of indigo and other pigments; they now show dirty stains and elaborate brushwork.

No. 54, 1887, 'at Hampstead Heath,' by the same artist, is, on the other hand, a well-preserved drawing, which has been quite recently acquired by the museum. The blues are particularly good, although the vegetable yellows may perhaps have faded somewhat in spite of the care which has been taken with the drawing.

John Cristall (1767-1821). No. 142, 1890. This recently-acquired drawing seems to have kept its hues well, and affords a good example of the style of colouring of the period and school to which it belongs.

Jacob Xavery (painted in 1757). No. 15, 1872. This has faded woefully. The sickly peaches and spectral grapes proclaim the evanescence of crimson lake, gamboge, and indigo.

Mary Moser, R.A. (1744-1819). No. 160, 1881. Very little is left of the original colour here: the 'roses and other flowers' are a complete wreck.

Francis Danby, A.R.A. (1793-1861). No. 480. The Prussian blue in the sky unchanged, but the pigments which once modified its hue have fled.

John Varley (1778-1842). No. 381. Hot iron reds show themselves in great force in clouds and elsewhere—even in the river; the modifying organic pigments with which they were mixed have nearly disappeared.

George Cattermole (1800-1868). No. 503. The crimson lake seems to have faded from the face in this drawing. In No. 507, painted in 1850, we have a good illustration of the warm brownish hues produced by the change of crimson lake.

Samuel Prout (1783-1852). Nos. 1,473, 1869, and 3,056, 1876, afford examples of the stability of true ultramarine. The skies in both these drawings are now quite out of harmony with the architectural features, some of the pigments in which must have faded. These drawings should be compared with others by the same artist which hang beside them, and in which the blues of the sky, as well as some of the pigments in other parts, have faded. Here it may be convenient to remark upon the startling prominence of the skies, and sometimes of the blue distances, in many water-colour drawings. The first glance on entering a room in which such works are gathered reveals the permanence of ultramarine and cobalt blue, the latter pigment being, of course, of comparatively recent introduction. But the want of harmony in such drawings furnishes evidence at the same time of the decay of many other pigments—of liquorice and tobacco juice, of yellow lake and brown pink, of indigo and rose pink.

William Henry Hunt (1790-1864). No. 1,031, 1873. Some of the ruddy hue from the cheeks of the boy appears to have gone. In 1,525, 1869, some of the pink in the apple-blossom has faded; the primrose-flowers are greener and less yellow than they once were, probably from a change in lemon yellow. The hawthorn-blossom in 1,470, 1869, has lost the faint rosy blush that was once visible in some of the flowers.

The above examples, selected almost at random, must suffice. But I may point to a different kind of injury, from which water-colour drawings sometimes suffer, by citing the case of W. Delamotte's 'View of Christ Church, Oxford.' This seems to have been pasted on wood, and to have been stained in consequence; at least, it appears likely that the brown spots in the sky may be traced to the mount.

A work by T. Barker (No. 134, 1878) and the 'Dieppe' of J. S. Cotman (No. 3,013, 1876) furnish additional examples of the same kind of damage.

CHAPTER XXV

CONSERVATION OF PICTURES AND DRAWINGS

It is generally conceded that a finished oil painting is best kept in such a position that it is exposed to daylight of just sufficient intensity for it to be well seen, the direct beams of the sun being excluded. In darkness, or even in approximate darkness, the lead-whites may tarnish, and the oil and resins darken. Even with the most moderate illumination, however, the more fugitive pigments, such as the cochineal and quercitron lakes, will in no long time alter and fade to such a serious extent as to destroy the 'keeping' of the work. But there is really no need to introduce these evanescent pigments, for every nuance the artist can desire may be produced with paints having a sufficient if not perfect degree of permanence. The question of artificial lighting here comes in. Against oil lamps, properly constructed and managed so as to avoid the production of smoke and soot, nothing can be urged; the same opinion may be given in reference to the use of electric incandescence lamps. The introduction of electric arc lamps seems less safe, even when the illumination they afford is reduced to the necessary minimum, for the light they emit is richer in those rays which, as a general rule, are peculiarly effective in bringing about chemical changes in the less stable pigments. But the light by which pictures are to be seen is but one of the conditions out of several which have to be considered in their conservation, though perhaps the most important. The mode of securing a picture in its frame is not an altogether trivial matter. In the majority of cases the expansion and contraction with variations of moisture and temperature, of panel and canvas, do not correspond accurately with the similar changes of the frame. In consequence, too great rigidity in the system of fixing adopted should be avoided. Blocks of indiarubber (*not* vulcanized), duly secured in the rebate, afford a convenient means of obtaining the necessary freedom of movement, while preventing the jar caused by accidental concussions. An equable temperature is another important condition; on no account should currents of hot or of cold air impinge directly upon the front or back of a painting. Moreover, this is not a mere question of temperature, for such currents of air may bring in particles of dust, and other impurities, while their hygroscopic condition is sure to vary. This question of moisture is of some moment. For if freshly-warmed air, which is pretty sure to be comparatively dry air, is allowed to come in contact with panels or canvases, it will withdraw from them some of their necessary hygroscopic moisture, and thus cause capricious and hurtful changes of size. Such changes, often repeated, cannot but result in the production of cracks and fissures in the oil paintings subjected to these varying conditions. The hygroscopic balance between picture and air can be maintained only by a due supply of moisture to the warmed air before the latter comes in contact with the painting; the warmer the air the more moisture must be added to it. The same reasoning applies to the entrance of cold air, but in this case, care must be taken that it is sufficiently dry not to deposit water upon the picture. For the purpose of regulating the hygroscopic condition of the atmosphere in a picture-gallery, the introduction of a dew-point thermometer is advisable. And there is another contrivance by means of which the presence of the right proportion of moisture in the air may be recognised. A strip of drawing-paper, another of primed canvas, and another of mahogany, all three being in a normal hygroscopic state, are to be separately balanced by means of counterpoises. When the air gets too dry, the strips will rise, owing to their loss of water; when excess of moisture is present, they will sink. So long as the equilibrium of the beams to the ends of which they are attached remains practically true, the air may be regarded as in a satisfactory hygroscopic condition. Three pairs of ordinary apothecaries' scales (or spring-balances) suffice for the construction of this apparatus, which should be protected by a glass case to which the air has free access. This question of the due amount of moisture in the air—neither in excess nor in defect—has scarcely received the attention it deserves. But it will be allowed on all hands that few conditions are more trying to pictures in oil or water-colour than those caused by currents of hot dry air rising directly below them during the day, succeeded by currents of cold moist air descending upon them down the surfaces of the walls at night.

The covering of an oil picture with glass, whatever objections may be urged against it from an artistic point of view, certainly secures the protection of its surface from the solid, and liquid and, to some extent, from the gaseous impurities in the air. But the back of pictures, especially of those painted on canvas, is often forgotten, yet excess of moisture and deleterious vapours and gases often enter from behind, and seriously discolour the painting-ground, and even the paint itself. Mention has previously been made of methods by which this cause of injury may be prevented by means of a double canvas, or a layer of white lead in powder, mixed with starch-water, applied to the back of the original canvas; American leather cloth, or even parchment-paper, affixed to the frame behind is nearly as effective.

A few words only are requisite as to the mounting and framing of water-colour paintings. On no account must the back of the paper on which a drawing is executed come into contact with any kind of wood, or even with an inferior sort of paper or mounting-cardboard. Injurious substances in the latter may travel forwards into the painting-ground, and affect the pigments, while wood may cause stains. Iron brads produce rust-spots. Flour-paste is not a sound material for mounting drawings; far better is the solution of indiarubber used by book-binders, and this may also be used for fixing to the back of the frame the sheet of paper which is there placed to exclude dust. If we could secure a water-colour drawing from dust, and yet allow of the escape of any water set free in the form of vapour when the drawing gets, from whatever cause, somewhat warmer than usual, we should have effected an improvement upon the ordinary plan of framing. In this, the moisture liberated from the paper and mount cannot escape, but condenses upon the glass when it cools, only to be reabsorbed by that surface of the paper which carries the pigments, where it favours chemical and physical changes, until the hygroscopic equilibrium of the whole system—frame, mount, lining, paper, etc.—is once more re-established. To hermetically seal a framed drawing, to the entire exclusion of all moisture and all air, is not possible. That under such conditions nearly every pigment would be unalterable has been long known. We should add to these observations upon the conservation of works in water colour that they should certainly be kept in a rather drier atmosphere than that recommended for oil paintings.

The plan of preserving the water-colour drawings of Turner, devised by Professor Ruskin, may be mentioned fitly here. It was described in a letter to the editor of the *Times* (October 28th, 1856). The recommendation is to enclose each work in a light wooden frame, under a glass, the interior surface of which is prevented from coming in contact with the drawing by means of a raised mount. A number of such frames (five to fifteen) are kept together in cases, which can be carried or wheeled to any part of the room where the drawings are to be studied. Each frame slides vertically into grooves in the case. Professor Ruskin's reasons include the following: 'A large number of the drawings are executed with body colour, the bloom of which any friction or handling would in a short period destroy.' This argument, it will be seen, is directed against the keeping of such works, in their unframed state, in portfolios. Another reason given by Mr. Ruskin is that in the case of these drawings 'their delicate tones of colour would be destroyed by continuous exposure to the light, or to smoke and dust.' He fortifies his position in reference to such exposure in a letter to the *Literary Gazette* (November 13th, 1858), in which he says that 'the officers of both the Louvre and of the British Museum, refuse to expose their best drawings or missal-pages to light, in consequence of ascertained damage received by such drawings as have been already exposed; and among the works of Turner I am prepared to name an example in which, the frame having protected a portion, while the rest was exposed, the covered portion is still rich and lovely in colour, while the exposed spaces are reduced in some parts nearly to white paper, and the colour in general to a dull brown.' 'That water-colours are not injured by darkness is also sufficiently proved by the exquisite preservation of missal paintings, when the books containing them have been but little used. Observe, then, you have simply this question to put to the public: "Will you have your Turner drawings to look at when you are at leisure, in a comfortable room, under such conditions as will preserve them to you for ever, or will you make an amusing exhibition of them (*if* amusing, which I doubt) for children and nurserymaids; dry your wet clothes by them, and shake off the dust from your feet upon them, for a score or two of years, and then send them to the waste-paper merchant?"' Mr. Ruskin in another letter to the *Times*, which appeared on October 21, 1859, wrote thus: 'I take no share in the responsibility of lighting the pictures either of Reynolds or Turner with gas; on the contrary, my experience would lead me to apprehend serious injury to those pictures from such a measure; and it is with profound regret that I have heard of its adoption.' Although considerable weight is rightly given to the opinions of Mr. Ruskin above quoted, it must not be forgotten that all paintings of the modern school are not to be classed with those of Turner and Reynolds, in respect to susceptibility to the injurious action of the products of the burning of gas and of continuous exposure to moderate light. When, therefore, Mr. Ruskin wrote (in the *Daily Telegraph*, July 5, 1876): 'I do not think it necessary to repeat my former statements respecting the injurious power of light on certain pigments rapidly, and on all eventually,' I find myself compelled to reject so sweeping an assertion. That light injures all pigments eventually cannot for one moment be conceded. And if we could but succeed in so modifying the light that illuminates our pictures as to remove from it certain particularly active beams, we might considerably augment the list of permanent pigments. Experiments on a small scale prove that several fluorescent substances, such as a solution of quinine sulphate, while intercepting dangerous rays, do not sensibly modify the colour of the light, and yet lessen its chemical activity. Possibly a transparent screen of this character will some day be used for our picture galleries. But the further discussion of this subject will be found in the next chapter, and to this I would refer my readers.

The question of the lighting of a gallery or room where pictures are to be displayed has been touched upon already at the beginning of the present chapter. We would now add that actual skylights are not without drawbacks. One of these, especially in the case of water colours, is the presence in large proportion in the light from the zenith of those rays which act most energetically upon pigments. A few observations as to gas cannot be excluded. Gas, before and after burning, is bad for pictures. The evil effects of an occasional escape of unburnt gas are less to be dreaded than those caused by the products of gaseous combustion. These products are sulphuric acid, sulphurous acid, carbonic acid, and the moisture which is formed at the same time. Thence results a hot, moist atmosphere laden with these corrosive compounds. The water vapour condenses into the liquid form and dissolves a part of the acids named above: the drops which trickle down are very injurious to paper, wood, canvas and pigments. In any case all the products of the combustion of gas should be removed from the room as they are formed. For even when there may be no visible condensation of liquid, the vapours formed are often absorbed as such by paper, canvas, etc., and do in that form their destructive work. An illustration of this fact is furnished by an analysis of the leather back of an old calf-bound volume. Owing to its absorption of the products of the burning of gas this back had decayed and fallen off, and was found to contain over 6 per cent. of free sulphuric acid.

With respect to the building itself in which pictures are to be kept, our aim should be to secure as far as possible pure air, an equable and agreeable but moderate temperature, and freedom from dust and dirt. Solidity of construction, a continuous damp-proof course, a certain degree of elevation above the ground-level, and double walls enclosing an air-space, are desirable, as conducing to uniformity of temperature, and preventing the condensation of moisture upon the interior of the rooms. Due provision should be made for excluding from the galleries themselves the dust and dirt which may be brought in by visitors. And it cannot be too strongly urged that the supply of fresh air, should not, as it were, accompany the visitors, but be brought in from an independent source. The place of in-take of such supply should not be gratings near to, or on the level of the ground, in out-of-the-way and dirty corners, the certain depositories of uncertain rubbish. From such sources air laden with organic and inorganic impurities can alone come. The question of the exclusion of fog and city-smoke may be mentioned here. Some kind of air-filter is useful. It is astonishing how effectively the solid and liquid particles suspended in a yellow fog may be strained off and intercepted by passing the air through a layer of loosely packed carded cotton—this air-filter of course requires occasional renewal. Moist white lead, that is, white lead in powder reduced to a paste by admixture with water, will absorb the sulphuretted hydrogen as well as the sulphuric and sulphurous acid present in town air. And if the walls of the galleries are coated with a distemper paint containing white lead, this absorption of impurities goes on continually. For these impurities are more readily absorbed by a properly-prepared distemper than by the changeable pigments in the pictures. To secure this result the distemper should be made, not with size, but with starch-water, just sufficiently strong to bind the particles together, and to the wall. For further particulars as to these and other arrangements for the conservation of pictures, especially in public galleries, the reader is referred to a paper on the subject in the *Portfolio* for 1882, pp. 106-108.

The conservation of pictures naturally leads us on to their restoration. Picture-restoration, like some other kinds of restoration, often involves injury, often renewal. It is frequently difficult, sometimes impossible, to re-establish the *status quo ante.* The operations involved should never be undertaken by the inexperienced amateur. And picture-restorers themselves are too often artists who have mistaken their profession, or who have been imperfectly trained. Many possess no power or appreciation of accurate draughtsmanship. Look, for illustrations, at those crucial parts, the hands and feet, in 'restored' pictures. Nor have they that exquisite sense of quality in colour and of delicate hues which will enable them to fill up properly actual gaps in a painted surface. Then the pigments they use are too often unsafe, and their vehicles unsatisfactory; so both soon alter. At the same time, the skill of some restorers in the matter of mechanical repairs—parquetting, transferring, relining, etc. —cannot be too highly extolled.

Several manuals of directions for restoring pictures have been given to the world; they are of very unequal value. Some of these tell you nothing, for they are intended merely to advertise the skill of the writers. Others advocate a treatment which may be called heroic, giving you solvents, not only for the discoloured varnish, but for glazings and paints. In reality, no directions can replace experience and skill. M. von Pettenkofer's method is one of the best known, but it is very rarely applicable with safety and success. The object of this method is the renewal of the transparency and continuity of the varnish by a process of re-solution *in situ*. With this intention, the picture is exposed in a closed shallow box to the action of the vapour evolved from moderately strong spirits of wine. This vapour dissolves the mastic on the surface of the picture, forming once more a spirit-mastic varnish. This, on exposure to the air, hardens, and leaves a shining coat of resin. But this resin, being necessarily discoloured and sinking into the cracks of the paint, makes them more prominent, while there is great danger of its being unequally distributed over the surface of the work.

When the varnish of an old picture is practically intact, but the surface is begrimed with soot and dirt, it should not be cleansed by the direct application of water, much less by the use of a solution of soap; but a loaf of household bread, not more than a day old, should be taken, and its crumb broken up into a tin canister furnished with a lid; it is important that no pieces of crust, and no fragments of crumb which have become hard by drying, should be introduced. Then the crumbs should be shaken out, in portions at a time, from the canister on to the varnished surface, and rolled gently thereon by means of the fingers. By repeating this operation until fresh crumbs no longer become discoloured, it is often possible to improve the appearance of a picture very greatly. In any case, it affords a useful preliminary to the removal of the old varnish where such a further step is imperatively required. Such removal is effected by the mechanical process of *chafing*. A single tear of pure mastic resin is ground or crushed to fine powder, and placed upon some unimportant part of the surface of the picture; but the operation *may* be begun without the aid of the mastic-powder. A gentle rotatory movement of the ends of the fingers soon reduces the old varnish-layer to powder, which is then removed by means of a soft badger-hair brush, or other suitable means. The work should be performed in a good light, and great care must be taken not to injure any tender glazings belonging to the painting itself. To ascertain whether the removal of the varnish has been carried far enough, the parts treated may be moistened with distilled water applied on a plug of carded cotton. When the operation is complete, and the surface is quite dry, a new coat of mastic-varnish may be applied, if possible in an artificially dried atmosphere. Sometimes a little dragon's blood, or other warm-coloured resin, is added to the mastic-varnish, in order to prevent the cold and raw look which a picture which has lost its old toned varnish frequently presents. An oil-painting in which no megilp has been used, and which has received, a year after completion, a thin layer of drying-oil containing a little copal-varnish, and then, after the lapse of a twelvemonth, its final coat of mastic-varnish, cannot be injured by the chafing process just described. And, even under less favourable conditions, it is the only method which can be recommended for general adoption. But it has its risks, and is not easily applicable in the case of pictures where the texture of a coarse canvas, or the grain of a panel, is conspicuously evident on the surface. To these remarks on the chafing process, we may add that it is sometimes advisable to re-varnish a picture with fresh mastic before commencing to remove the old; a day or two afterwards both layers may be removed together.

The removal of old varnish by the use of solvents is a hazardous, though easy, operation. The liquid usually employed for this purpose is spirits of wine, of about 60° overproof, diluted with one-fourth of its bulk of distilled water. It is applied by means of plugs of carded cotton, the action of the solvent being arrested, when necessary, by instantly moistening the spot with spirits of turpentine on another plug, or, in some cases, with linseed-oil. But when mastic megilp has been used as the painting-medium, it also, as well as the pigments associated with it, may be removed by treatment with these solvents. And it must be remembered that some artists introduce layers or touches of water-colours in their oil-pictures; these are almost certain to be affected by spirits of wine. Sometimes further injury to them may be arrested by the application of linseed-oil. Whenever a solvent is used in cleaning a picture, the cotton tufts employed should be examined carefully from time to time, in order to see that no actual pigment has coloured them—that they are stained by nothing but the brown varnish. Other solvents besides those named are sometimes used in cleaning pictures, particularly where hard or oily varnishes have to be removed. Such solvents are acetone, benzene, chloroform, and solutions of caustic alkalies. Great risk of injury attends their employment—indeed, the application of any kind of solvent is fraught with danger, and should never be attempted by the inexperienced.

The usual plan of filling up actual gaps in the priming or gesso-grounds of old pictures is by means of plaster of Paris. When this has set, its surface is levelled by gentle attrition with a cork and dry whitening, or cuttle-fish. Great care is needed in order to prevent the pigments surrounding the place from being abraded. I have found asbestos-putty to be an excellent substitute for plaster in many cases; its surface is made smooth and level by means of a painting palette-knife. But it cannot be tinted with water, or tempera-colours, in order to make it match the hues of the neighbouring parts of the picture; oil-colours must be used. In any necessary replacements of lost colours in old oil-paintings, it has been recommended to use not oil-colours, but water-colours, as these can always be distinguished from the old work, and, if necessary, removed; this can be done on 'stoppings' of whitening and size, as well as on those of plaster. Tempera-pictures should, I think, be repaired with dry pigments mixed with egg-yolk, as, in this case, when the final varnish is applied, a general harmony of effect is produced. If water-colours are introduced in repairing an oil-painting, the coat of varnish subsequently added is sure to deepen and darken them, unless this change has been allowed for during the progress of the work, by no means an easy thing to manage. When in any kind of re-painting oil-paints are used, they should be mixed stiffly with a very little copal-varnish and spirits of turpentine, and should be rather lighter and less yellow in tone than the colours they are intended to match, since darkening and yellowing, in some degree, however slight, are sure to occur subsequently.

The cleansing and restoration of paintings executed in fresco require special care. Additions to supply colour which has scaled off are best made in tempera. When a fresco has become grimy by exposure to the smoky air of a city methylated spirits of wine, applied freely on tufts of carded cotton, removes the tarry and sooty impurities which a previous careful brushing of the painted surface has failed to dislodge. Attempts to clear the clouded portions of an injured fresco by means of distilled water or aërated distilled water are usually attended with but slight success. The films which obscure the surface in such cases sometimes consist of sulphate of lime, produced from the carbonate of lime present by the action of the sulphuric acid occurring in the atmosphere of places where gas and coal are burnt.

In getting rid of this somewhat opaque film by means of water portions of the pigment are commonly removed. When a fresco has been dusted and then cleansed with spirits of wine it should be allowed to dry thoroughly, and then be treated with the paraffin-copal medium (described on p. 109): this medium must, however, be largely diluted with spirits of turpentine before applying it to the fresco. By this treatment the dead or matt surface of the painting is preserved, the obscuring films are rendered translucent, and the picture may, when cleansing is again required, be safely sponged with pure water or with weak spirits of wine.

CHAPTER XXVI

TRIALS OF PIGMENTS

THE testing of pigments for genuineness and for purity has been discussed incidentally in Chapters XIII. to XIX. of the present volume. Though genuineness and purity* have ofteu an important bearing upon the question of permanence, this last quality must be ascertained by independent experiments of another order. The study of old paintings often furnishes us with results of some value, the results of unintentional testings. But for the purpose of securing complete and wholly trustworthy information, we must know precisely all the materials and all the conditions of our trials. Not only must the painting grounds, the mediums, and the pigments, be chemically examined, but we must be in a position to state the character of the atmosphere in which they have been exposed, and the nature and amount of the solar, or other radiations, to which they have been subjected. In the great majority of these trials accurate data as to materials and conditions are wholly wanting; even the South Kensington report affords us no information as to the composition of the pigments employed, nothing

* The chromatic values of pigments—their approach in hue, etc., to recognised standards of excellence—are not here taken into account. Such data may be obtained by the use of Lovibond's tintometer, when once the chromatic elements of a pigment in terms of the degrees of the standard glasses employed in this instrument have been determined.

more than their commercial names, so that we have to take on trust their genuineness and purity. However, in this same most important series, quite unusual, if not unprecedented, care was taken in order to determine the conditions, physical and chemical, under which the pigments were tested. In my own experiments, carried on between 1856 and 1879, in somewhat desultory fashion, and extended and improved since 1880, the influence of purity of sample, of the presence of moisture and of oxygen, and of the nature of the 'light,' has not been neglected; the full details of the methods adopted, and of the results obtained, could not be appropriately introduced into an elementary manual; mention will be made of the chief conclusions reached in the present chapter, there are also numerous references to them in Chapters XIII. to XXII.

In many early treatises on painting we find observations as to the degrees of stability possessed by various pigments, along with suggestions as to methods of treatment by which in certain cases greater permanency may be attained. To some of these observations and suggestions reference has been made in those chapters of the present volume which deal with pigments; many of the remainder are now without practical importance, referring as they do to pigments which have been rightly discarded.

It is only of recent years that trials of pigments have been made with any approach to exactness. But in the majority of cases no information has been secured concerning the purity and genuineness of the pigments with which the experiments have been made. I have not been able to find a single chemical analysis of any one of the pigments tested. The chromatic analysis of the light they severally reflect has, however, been recorded in the case of the water-colour paints examined by Dr. Russell and Captain Abney, who have likewise furnished some particulars as to the intensity of the solar radiations to which the pigments were subjected.

More than a century ago Sir Joshua Reynolds tested, in a rough way, the stability of some of the paints he employed. Two causes detract from the value of his results. In the first place, the information furnished concerning the nature of many of the pigments he tried is too imperfect to be of any use; secondly, we are not acquainted with the conditions under which his specimens have been kept during the whole period since they left his hands. Two of his trial canvases are preserved in the Royal Academy. Although the darkening and embrowning of the surface are general, and though the names of the pigments employed are often undecipherable or meaningless, yet something useful remains. From the experiments made in 1772 we may gather the following facts. The proper hues of several pigments remain in a measure: orpiment, or king's yellow in crystals; yellow lake, with wax and drying oil; gamboge, with lake and Venice turpentine; gamboge, with (Venice) turpentine; prepared gamboge, with wax; and verditer, with varnish. On the other hand, gamboge with oil, lake with oil, and many other pigments of organic origin, when unmixed with wax or varnish, are names only, or, at the most, brown discolourations.

M. J. Blockz, in his 'Peinture à l'Huile,' gives the results of a number of experiments made by M. J. Dyckman. He condemns, for various reasons, not only asphalt, but also terre verte, cobalt green, emerald green, raw sienna, raw umber, ivory brown, and all burnt madders. Cassel earth was slightly changed; brown ochre, burnt sienna, Mars brown, ivory black, and vine black, proved to be permanent. His lists of incompatible pigments are somewhat unnecessarily extended, not being justified, in all particulars, by further and more careful experiments.

The experiments of Mr. F. W. Andrew, of the South Kensington Museum, have been carried on since 1876, but have been confined to water-colours, both moist and cake. His chief results will be found recorded along with my own, in Part III., in the paragraphs devoted to the consideration of the several pigments. His water-colour washes, generally spread on Whatman paper, or Whatman board, were exposed for periods varying from twenty-eight to eighty-two months, in a south window, to all the sunshine available; half of each wash was doubled back, and so far excluded from light. In some cases a third part of the coloured slip was exposed to the air and light without the protection of glass. The chief paints which were unaffected, at all events, so far as some specimens were concerned, by the exposure, are included in the following list: yellow ochre, raw sienna, deep cadmium, Mars red, light red, Indian red, oxide of chromium, Leitch's blue, cobalt, artificial ultramarine, Prussian blue, raw umber, burnt umber. Naples yellow (true) became blackish in darkness, but was unaltered in light, and orange chrome showed dark patches. Further details must be given as to the pigments which were affected. In the tabular statements appended a selection from the results afforded by eleven sets of experiments is recorded, the letters *C* and *M* prefixed to the entries respectively denoting the employment of cake or moist colours. The numerical values representing the residual hues are rough approximations only, but in some instances they were determined by means of comparisons with standard-coloured liquids, contained in glass cylinders graduated into ten equal parts. The tintometer of Mr. J. W. Lovibond, of Salisbury, may be employed with advantage in these estimations.

CHANGES IN WATER-COLOURS

Name of Pigment	*Months of Exposure*	*Residual Depth (Original = 10)*	*Residual Hue, and Remarks*
M. Aureolin	28	10	Verges on orange-yellow.
M. Aureolin	62	10	
M. Aureolin	82	9	
C. Gamboge	28	4	
M. Gamboge	28	9	
M. Gamboge	82	7	
M. Yellow lake	62	2	Yellowish grey.
M. Yellow lake	82	0	
C. Yellow madder	28	1	
M. Yellow madder	28	3	Pinkish grey.
M. Yellow madder	62	1	
C. Indian yellow	28	8, 9	
M. Indian yellow	28	9	
M. Indian yellow	60	7	
M. Indian yellow	82	6	
M. Pale cadmium	82	1	Pale buff.
M. Vermilion	62	—	Blackish.
M. Vermilion	82	—	Black.
M. Carmine	28	0	Pale grey.
M. Crimson lake	28	0	Greenish grey.
M. Burnt carmine	28	1	,,
C. Pink madder	28	1	
M. Rose madder	28	3	
M. Rose madder	62	1	Pinkish grey.
M. Rose madder	82	1	,,
M. Madder carmine	62	2	
M. Madder carmine	82	0	
M. Purple madder	28	4	A pale wash.
M. Purple madder	62	7	
M. Purple madder	82	2	
C. Brown madder	28	1	Warm grey.
M. Brown madder	28	1	,,
M. Brown madder	82	0	
C. Indigo	28	5	Greenish grey.
M. Indigo	62	1	
M. Raw umber	60	9	Rather yellower.
M. Vandyke brown	28	7	
M. Vandyke brown	60	1	
M. Bone brown	62	8	

The pigments containing lead, such as the ordinary chromes, and those having a copper basis, like emerald green, had altered capriciously, losing part of their original colour, and becoming tarnished or embrowned in patches. Brown pink faded like the yellow lakes, but acquired a bluish-grey residual hue.

The testing of the madder colours is so important that I introduce here a few additional experiments selected from my own note-books. The washes of the moist-colour paints were, as far as possible, of the same depth of tone, and they were all exposed together in a glazed frame to one year's sunshine:

Name of Pigment	*Residual Depth (Original = 10)*	*Change of Hue, etc.*
1. Rose madder	8	Slightly more purplish.
2. Madder carmine	10	Almost unchanged.
3. Madder carmine	8	Much more purplish. This sample was from another source.
4. Madder red	6	Less red, more purplish.
5. Purple madder	7	Duller, less red, more blue.
6. Brown madder	8	Less red, more yellow-brown.

In contrast to the above results with madder carmine, the following experiment with the ordinary carmine (prepared from cochineal) is instructive. On a sheet of Whatman paper, a space of 10 inches in length by 4 inches in width was covered with a uniform wash of the moist paint, having a depth of tint about equal to that of the petals of the old China rose. This coloured strip was then subjected to summer sunshine in such a way that successive single inches of its length received the light (during the same hours of similarly bright days) for periods of 2, 4, 8, 12, 20, 26, 30, 40, and 100 hours, one single inch at one end being, however, protected completely from all access of light. The exposure of 100 hours sufficed to bleach the last breadth completely, but had the rate of fading been in a simple arithmetical progression, a much shorter exposure would have sufficed. In fact, the bleaching action was far more energetic during the first period of 2 hours than during the second, about 20 per cent. of the original colour having been destroyed during these two first hours, while during the second equal period the loss of depth did not exceed one-tenth of this amount. Moreover, it was noticed that the change of *hue* consequent upon the first exposure was different in kind to that which occurred subsequently.

Professor O. N. Rood's experiments.—In his 'Modern Chromatics,' pages 90 and 91, Professor Rood gives the results of a few trials which he made as to the effect on washes of water-colours laid on ordinary drawing-paper, of three and a half months' exposure to summer sunlight.

These pigments were unaffected:

Cadmium yellow, yellow ochre, Roman ochre.
Indian red, light red, Jaune de Mars.
Terre verte.
Cobalt, French blue, smalt.
Burnt umber, burnt sienna.

The following pigments were all affected. The sequence represents the amount of alteration, the list commencing with those colours which suffered but little change:

Name of Pigment	*Nature of Change*	*Name of Pigment*	*Nature of Change*
1. Chrome yellow -	Slightly greenish.	13. Hooker's green -	More bluish.
2. Red lead - -	Less orange.	14. Gamboge - - -	Fades, greyish.
3. Naples yellow -	Slightly greenish brown.	15. Bistre - - - -	Fades, greyish.
		16. Brown madder -	Fades.
4. Raw sienna - -	Fades, yellower.	17. Neutral tint - -	Fades.
5. Vermilion - -	Darkens, brownish.	18. Vandyke brown -	Fades, greyer.
6. Aureolin - - -	Fades slightly.	19. Indigo - - - -	Fades.
7. Indian yellow -	Fades slightly.	20. Brown pink - -	Fades greatly.
8. Antwerp blue -	Fades slightly.	21. Violet carmine -	Fades greatly, brownish.
9. Emerald green -	Fades slightly.		
10. Rose madder	Fades slightly, purplish.	22. Yellow lake - -	Fades greatly, brownish.
11. Sepia - - - -	Fades slightly.	23. Crimson lake - -	Fades out.
12. Prussian blue -	Fades somewhat.	24. Carmine - - -	Fades out.

Professor Rood adds that rose madder, brown madder, and purple madder were all a little affected by an exposure to sunshine for seventy hours, and that *pale* washes were completely obliterated by a much shorter exposure to sunshine in the case of carmine, dragon's blood, yellow lake, gall-stone, brown pink, Italian pink, and violet carmine.

Professor W. N. Hartley's Experiments.—On September 4, 1886, Professor Hartley read, before the British Association at Birmingham, a paper on 'The Fading of Water Colours.' His trials as to the effect on pigments of a comparatively brief exposure to intermittent sunshine in pure air, may be thus summarized. Washes on the best drawing-paper were the subject of the experiment:

Gamboge.—Pale washes were completely bleached in three days; in a week strong washes were much lightened in colour, and rendered dull, even three hours' exposure producing a very visible effect.

Crimson lake.—Six hours' exposure to sunlight and air almost bleaches pale washes, while three days or eighteen to twenty-four hours of intermittent sunshine cause dark crimson tones to become very much lighter, the hue of the pigment being altered.

Light red, Indian red, and vermilion were unaffected.

Olive green and brown pink were rendered lighter in colour by six hours' exposure, the former pigment becoming bluish and the latter brownish in hue.

Indigo, cobalt, and artificial ultramarine were unaffected.

Brown madder became rather lighter after eight days' or forty-eight hours' exposure.

Bistre faded with great rapidity, a light wash appearing much paler after six hours.

Sepia.—A pale wash became colder in hue, but not very perceptibly paler.

In a second series of experiments, sectors of paper discs, washed with various pigments, were enclosed between glass-plates, the edges of which were fastened with gummed paper. Under these circumstances, crimson lake and bistre were found to have been considerably altered by five hours' exposure—somewhat more so, indeed, than was the case when these pigments were freely exposed to the air.

All the results above noted are in practical accord with those obtained by other observers. The exposure to intermittent sunshine 'for six hours a day during fourteen days,' does not produce a sensible effect upon vermilion and indigo. Had Professor Hartley extended his observations a few weeks longer, his conclusions as to these pigments must have agreed with those which we have given, and, therefore, with the unanimous verdict of all other scientific observers. His statement that 'indigo is permanent' (Brit. Assoc. Report, 1886, p. 581) must, therefore, be modified into, 'indigo appears to have suffered no change after fourteen days' exposure to intermittent sunshine' A similar alteration is demanded with regard to the stability of vermilion.

Professor Hartley's experiments with water-colour washes on paper enclosed between glasses, require a few words of comment. He is clearly and rightly dissatisfied with this method of trial. A supply of atmospheric oxygen, and of hygroscopic moisture, amply sufficient for large chemical alteration and oxidation of the enclosed pigments, was certainly present. And the glasses *did* accelerate the action, not because of 'the very slight tint of the plate-glass,' but in spite of it. This acceleration of change is mainly caused by the continued presence of moisture in the confined space between the two glasses—it cannot escape as it freely, and to a very great extent, does escape, when a piece of tinted paper is exposed to sunshine in free air. I showed, indeed, in my lectures at the Royal Academy, so long ago as 1880, that the fading of many fugitive pigments is greatly lessened, when not altogether prevented, by enclosing the paper washed with them in a glass tube, the air of which is kept dry by means of some strongly hygroscopic substance. When both moisture and air are excluded (using a sealed vacuous tube), the suspension of fading and alteration of hue is still more marked and general.

It should be added here that Professor Hartley found that cadmium yellow and Indian yellow are bleached by peroxide of hydrogen, and changed into a muddy yellow by sulphurous acid. This reagent bleaches artificial ultramarine, and dulls vermilion. He attributes the partial or complete destruction of the blue component of the hues in certain old drawings which have been long exposed to air and light, to the presence of acids or acid substances in the air, in the paper, or in the red ferruginous pigments with which the blue colouring substances in question have been associated. These blue pigments could have been nothing other than Prussian blue, indigo, or natural ultramarine. I have ascertained, by direct experiments on old drawings, that the latter was but rarely employed for mixed tints, but it is quite probable that the reds prepared from colcothar, with which it may have been occasionally mingled, would sometimes contain enough acid salts (certain ferric sulphates) to destroy its colour. The products of the burning of gas and of coal would also be rich enough in sulphuric acid to produce the same effect. I am unable to endorse Professor Hartley's statement that the best drawing-papers have an acid reaction. The results of numerous trials of all kinds of drawing-papers, good and bad, have been already given in Chapter I., and so a mere reference to the facts, as derived from my own observations, must here suffice. I find that these sized papers are generally perfectly neutral to test-papers, and that inferior papers are more often slightly alkaline than acid.

Mr. W. Simpson's Experiments.—Some washes of water-colour, of thirty-one different kinds, were made upon cards by Mr. W. Simpson. He so cut the cards as to divide each coloured strip in half; one section was preserved in darkness, the other was exposed in an eastern aspect on the shutter of a house in London for fifteen years, but the sun did not shine upon the specimens after ten o'clock in the morning. As they were not tightly framed, the cards became a good deal discoloured by the absorption of noxious vapours and dirt. The results were:

Name of Pigment	*Nature of Change*	*Name of Pigment*	*Nature of Change*
Yellow ochre	None.	Madder lake	More purplish.
Indian yellow	Faded considerably.	Purple madder	Hue altered.
Lemon yellow	None perceptible.	*Brown madder	Loss of redness.
'Newman's permanent yellow'	None.	Emerald green	Slight.
Cadmium yellow	Perhaps browner.	Cyanine blue	Apparently none.
Chrome yellow	Faded considerably.	*Prussian blue	None.
Brown pink	Faded.	French blue	Faded very slightly.
*Vermilion	None.	Cobalt	None.
Light red	None.	Ultramarine	None.
Indian red	None.	Indigo, rather deep	Very pale grey.
Crimson lake	Gone.	Burnt sienna	None.
Carmine	Gone.	*Vandyke brown	None.
		*Sepia	Faded very slightly.
		*Bistre	None.

It will be noted that the above results are for the most part in agreement with those recorded by other experimenters; the chief exceptions are marked with a star. Vermilion is usually blackened, but it is possible that the sample employed in these experiments was the less changeable native form or cinnabar. The Vandyke brown, too, was probably the earthy rather than the bituminous variety; the slightness of the change recorded for madder brown and sepia, and the absence of any alteration on the part of bistre, are less easy of explanation. The madder pigments seem to have stood more than usually well, but they often exhibit large differences of stability. Nor must it be forgotten, in assigning values to the above results, that this trial of fifteen years' exposure was not of the severest kind. Although, on the one hand, there was the imperfect exclusion of an injurious London atmosphere, on the other hand, the energy of the solar radiation was much reduced by the prevalent condition of the smoky air, while the intermittent and capricious sunshine of the Metropolis never fell on the trial cards after ten a.m.

The late Mr. R. H. Soden-Smith, of the South Kensington Museum, kindly placed at my disposal a large number of specimens of old water-colour cakes and of powder colours intended for oil-painting. One set consists now of eleven cakes or fragments of cakes (in their original box) bought about the year 1815 of Newman, in Soho Square. This set is peculiarly interesting as the colours, which all bear the name of the maker and his device, represent those used by many of the best English water-colour painters during the first quarter of the present century. The cakes are: Indian yellow, raw sienna, raw umber, burnt sienna, burnt umber, vermilion, carmine, burnt carmine, pink madder, ultramarine, indigo; neutral tint and sepia are missing. On comparing the hues of the first nine of these paints and of the indigo with the hues of the corresponding cake-colours as sold by the same house in 1886, no appreciable differences were detected save in the case of the raw umber. Here the pigment of 1815 showed a more beautiful nuance than that of 1886. On making comparative tests of the stability, under exposure to sunshine, of the two sets of pigments, the results were found to be practically identical. One cannot, therefore, claim for the water-colour paints in use seventy-five years ago a degree of permanence greater than that possessed by their representatives of to-day.

By far the most important series of trials of water-colour paints yet published is that to be found in the report by Dr. Russell and Captain Abney to the Science and Art Department (1888). The reporters endeavoured to give precision to their experiments and their conclusions by a careful comparison of the effective radiation from different sources of light. The first part of their report is in part occupied by a discussion of the relative values of direct sunlight, light from clouds, and from an overcast or clear sky, and light from artificial sources. Several cognate subjects are also discussed therein, such as the number of years of exposure which pigments would require, if in the picture-galleries of South Kensington, in order that they might suffer the same changes as those caused by three or twenty-two months' exposure in a southern aspect outside the Museum. Part II. of the report contains the results of twelve sets of experiments with various pigments. In all the series the same paper (Whatman's) was used. In the paragraph relating to this subject there is, however, one curious error, and one obscure statement (page 27). It is quite impossible that the paper used—its weight per ream is not given—could have contained so little as nearly 1 grain of sizing matter per square foot; 10 grains is a more probable quantity. The next following sentences do not state the condition of the papers which absorbed from a moist atmosphere from 12·07 to 12·46 per cent. 'of their weight of water.' Were they dried previously, and, if so, at what temperature? We ought to have been told within what limits the percentage of water in these papers varied during the course of the trials. I have pointed out for many years past the importance of this hygroscopic moisture in paper in reference to the fading of pigments. Eight tints of each pigment were applied to strips of paper 8 inches long by 2 inches wide; they were exposed in tubes open at both ends, but having the upper extremity curved downwards so as to exclude wet and dirt. Of course exposure on a wall facing nearly south constituted a very severe test, yet the circulation of air in the tubes was more advantageous to the pigments than would have been the steamy heat of a closed vessel, or even of an ordinary paper-backed picture-frame. But on the other hand, this arrangement allowed the free access to the pigments of any noxious gases, such as sulphurous and sulphuric acids, and sulphuretted hydrogen, which might have been at any time present in the atmosphere. The general results of this first series of trials are gathered in the following table, the exposure in all cases lasting from May, 1886, until March, 1888. The pigments are arranged in the order of instability, the most fugitive being placed first:

		Show no change.
*Carmine.	Cadmium yellow.	Yellow ochre.
*Crimson lake.	Leitch's blue.	Chrome yellow.
*Purple madder.	*Violet carmine.	Lemon yellow.
*Scarlet lake.	*Purple carmine.	Raw sienna.
*Payne's grey.	*Sepia.	Indian red.
*Naples yellow.	Aureolin.	Venetian red.
*Olive green.	Rose madder.	Burnt sienna.
*Indigo.	Permanent blue.	Terre verte.
*Brown madder.	Antwerp blue.	Chromium oxide.
*Gamboge.	Madder lake.	Prussian blue.
*Vandyke brown.	Vermilion.	Cobalt.
*Brown pink.	Emerald green.	French blue.
*Indian yellow.	Burnt umber.	Ultramarine ash.

The pigments marked with an asterisk were found to have distinctly altered either in depth or hue by a much shorter exposure, from May to August 14, 1886.

In a second series of trials the tinted papers were dried, and then introduced into the tubes, which had been previously heated; the specimens were then sealed up hermetically: as no moisture-absorbing material was enclosed with the papers, traces of water must have been present. My own much earlier results were abundantly confirmed by those obtained in this series, for the number of pigments which proved to be permanent under these conditions was double that of the first series. Brown madder and Prussian blue were, however, acted upon in this second series. Dr. Russell and Captain Abney make the remark that of the eight colours which remained unchanged in dry air, but were acted on in ordinary air, all, with the single exception of madder lake, are mineral colours. But this is not correct, for the pigments named are—madder lake, olive green, Payne's grey, sepia, Naples yellow, cadmium yellow, emerald green, and burnt umber, and of these the first four are either wholly or partly of organic origin, while the seventh contains an acetate.

In the next series of experiments, the pigments were exposed in the presence of moisture-laden air. Very few colours withstood this test—none of organic origin; both Prussian blue and Antwerp blue were entirely destroyed.

An atmosphere of moist hydrogen gas was employed in the fourth series. Under these conditions carmine, crimson lake, madder lake, brown madder, olive green, indigo, Payne's grey, sepia, and Vandyke brown, suffered no change.

When, as in the fifth series, both moisture and oxygen were excluded, scarcely any even of the most fugitive pigments were affected. Vermilion, however, as in all the other experiments, became black. We know that the reason for the change is physical, not chemical.

In the sixth series it was proved that the addition of ox-gall had no beneficial effect in lessening the change of hue and tone in fugitive pigments.

The remaining series, save the twelfth and last, were devised in order to learn what influence upon the stability of pigments might be exerted by admixture with Chinese white, by exposure to the light of the electric arc, by heat without light, by heat and light together, and by exposure to the light transmitted through coloured glasses. Amongst the results recorded, we may note the decided changes in several pigments caused by admixture with Chinese white, and by heating the prepared paper slips in sealed tubes for seven hours a day for three weeks, all light being excluded. In a twelfth series of trials, the pigments were exposed in a picture-frame under glass in such conditions, and to such an amount of light, as might be taken to represent the ordinary circumstances in which pictures are kept. The frame was exposed from August 6, 1886, until May 6, 1888, to very bright light, but not to sunshine. Gamboge, indigo, Naples yellow, brown pink, carmine, and Vandyke brown, had faded in varying degrees. Some remarks on these results will be found further on in the present chapter; they are of extreme importance, considering the large use that has been made of these pigments by our water-colourists, and the mild treatment to which they were subjected during the short period of twenty-one months.

For the results obtained with mixtures of pigments under varied conditions of exposure, we must refer our readers to the report itself. It may, however, be remarked that, in the great majority of cases, the changes of tone and hue which occurred were such as might have been predicted from the known degrees of stability of the several constituents of the mixtures. Here, as elsewhere in the report, we find frequent mention of the strange, but long known, recovery in darkness of its colour by Prussian blue which has been bleached by sunlight.

In the fourth appendix to this report is an instructive list of the pigments employed by some of the most distinguished artists using water-colours. Forty-six painters replied to the invitation of the Science and Art Department; from their answers we learn that a large proportion of them include in their palettes many pigments which must be unhesitatingly condemned on account of their want of stability. Thus no less than 17 out the 46 artists who responded to the appeal employ three of the most fugitive pigments in the series—namely, gamboge, brown madder, and indigo. Converted into percentages, we may say that 37 out of 100 painters in water-colours use these three untrustworthy pigments, besides others which are worse, and others which are little better; of course, they employ also certain colours as to the stability of which there is no question. The following tabular statement gives the proportion of artists, per 100, who use the eleven perishable pigments named below:

Gamboge	70	Faded to 7.	1 represents the lightest tint, 8 the darkest. The washes of pigment were fully exposed for twenty-two months.
Indian yellow	24	Faded to 6.	
Vermilion	70	Gone black.	
Carmine	8	Gone.	
Crimson lake	22	Gone.	
Purple madder	28	Faded to 8.	
Brown madder	74	Faded to 3.	
Brown pink	11	Faded to 7.	
Vandyke brown	74	Gone.	
Sepia	65	Faded to 8.	
Indigo	52	Faded to 8.	

We call these pigments perishable with good reason. For, according to the report under review, all of them were found to have faded, materially and conspicuously, after twenty-two months' full exposure in a south aspect, while three of them had entirely disappeared, and another (vermilion) had become black. But this is not all. For under a less severe trial (p. 45)—namely, exposure for the same time, not to direct sunlight, but to a very bright light from a window, 'under conditions approximating to those to which pictures are usually subjected'—six out of the eleven pigments had faded, though in varying degrees. With these figures and results before us, it is impossible to resist the conclusion that a life of 100 years is too much to allow to many of the water-colour drawings of the present day. What shall we say, then, as to the stability of the works of the earlier masters of the English water-colour school? How much care in the exclusion of 'the more fugitive colours' was taken by the water-colourists of 1780 to 1850? Could it be honestly said of any large number of such works, in which gamboge, brown pink, crimson lake, sap-green, indigo, and sepia, were generally employed without stint, 'that about a century of exposure would have to be given to water-colour drawings in galleries lighted as are those at South Kensington before any marked deterioration would be visible in them'?* (Report, p. 46.)

Dr. Russell and Captain Abney add, indeed, the proviso, 'If painted with any but the more fugitive colours.' But this condition cannot be said to have been fulfilled by the works in question; for in the great majority of them, most of the six fugitive pigments which we have just named were freely employed. And it is these very pigments which have

* In the preceding brief *résumé* of certain parts of the South Kensington Report no reference has been made to an argument, developed in §§ x. to xv., in which it is contended that 'if a certain tint be exposed to an intensity of radiation which we will call 100, and which bleaches it in, say, 1 hour, then, if a similar tint be exposed to an intensity 1, it will require 100 hours' exposure to it to effect the same bleaching.' The universal applicability of this conclusion cannot be conceded by those who are familiar with numerous instances in which no chemical or physical change occurs when certain substances are exposed continuously for long periods to a particular temperature, yet, when they are heated but a degree or two higher, instantly alter, decompose, or react, as the case may be.

been proved by the reporters themselves to suffer 'marked deterioration' by an exposure of twenty-one months only to strong daylight without direct sunshine. Moreover, it must not be forgotten that the fading of a single important pigment in a water-colour drawing is ruinous to the whole effect, destroying the balance of the chromatic scheme of the artist more effectually than a slight, but equal, degradation of all the hues.

Amongst the series of trials of oil-paints made by the author of this handbook, one set first arranged in 1880 may be described here. Chance's colourless plate-glass was employed as the painting-ground, so as to avoid all interference with the pigments from the surface on which they were spread; glass presents the further advantage of permitting a complete examination of the back of each specimen, and of changes in its translucency, opacity, or texture. Each glass measured 8 inches by 6; the complete series was prepared in duplicate—one for preservation in darkness, the other for exposure to all the light that could be secured (in Kew) during five years in a window facing nearly south-west. The majority of the paints tried were obtained from four firms (Messrs. Winsor and Newton, Messrs. Roberson and Co., M. Edouard of Paris, and Schoenfeld of Düsseldorf). Specimens of each pigment were reserved for further examination and analysis. Some of the chief results obtained are given in the annexed table; some remarks on the changes observed in some of the pigments which had been mixed with flake white are added:

Pigment	*Years of Exposure*	*Residual Depth* (*Original* = 10)	*Change of Hue, and Remarks*
Yellow ochre	5	10	Browner; more translucent.
Auréolin	5	9	None.
Indian yellow	5	8	Slightly brownish.
Naples yellow (true)	5	10	None.

Pigment	*Years of Exposure*	*Residual Depth* (*Original*=10)	*Change of Hue, and Remarks*
Pale yellow madder	2	7	Greyish salmon when mixed with flake white.
Deep yellow madder	2	6	Dirty pink when mixed with flake white.
Laque Brun Jaune	2	7	Lost much yellow.
Laque Brun Foncé	2	8	Lost much yellow.
Laque Robert, No. 5	2	2	Warm grey when mixed with flake white.
Laque Robert, No. 6	2	4	Warm grey when mixed with flake white.
Scarlet lake	5	7	Dull pinkish red.
Crimson lake	5	1	Almost gone.
Madder red	2	10	None.
Madder carmine	5	9·5	None.
Madder brown	2	9	Rather duller.
Prussian blue	5	8·5	Slightly greener.
Indigo	5	8	Slightly greener.
Artificial ultramarine	5	10	None.

I devote the few pages remaining at my disposal to a subject connected very intimately, not only with trials of pigments, but also with the conservation of paintings. A long and somewhat heated discussion as to the effect of light upon water-colour drawings was commenced by Mr. (now Sir) J. C. Robinson by a letter which appeared in the *Times* of March 11, 1886. The writer affirmed, with regard to the collection of English water-colour drawings at the South Kensington Museum, that, having 'been continuously exhibited in the full daylight for twenty or thirty years past, by the mere fact of such exposure all these drawings have been more or less irrevocably injured, and that, in many cases, the specimens are now, as it were, but the pale ghosts of their former selves. These treasures, then, have sufficed for the delectation of one generation only, and we ourselves have practically used them up.' Mr. Robinson adds his recommendation that all water-colour drawings which ought to be fully and freely shown to the public, should be so shown only in the evening, and by the aid of the electric light.

To this letter Sir J. D. Linton replied in the *Times* of March 18th, traversing the statement with regard to the condition of the South Kensington drawings, and affirming that both the older tinted works, and those painted in 'pure colour' are slightly lowered in tone, but in other respects very little changed. Moreover, he cites one drawing by Turner ('Warkworth Castle,' No. 547), and the drawings by W. Havell, as 'absolutely richer and deeper in tone than when they were first painted.' He adds that the modern work, such as the 'Hawthorn Blossom' and two other drawings by William H. Hunt, is 'as brilliant' as on the day when painted: he speaks similarly of the 'Nottingham' by De Wint, the 'Windsor Castle' and 'Windermere,' by David Cox, and of all the drawings by Cattermole. Sir J. D. Linton considers that Mr. Robinson has fallen 'into the common error of confusing the so-called fading of colour by the action of light with the chemical changes chiefly caused by the unfortunate use of injudicious colours, particularly of Indian red, which in the skies of some of the drawings of the masters has had a deleterious effect.' He concludes by saying that he writes in order 'to combat the popular idea that water-colours fade by the action of light.'

The third contribution to the discussion was furnished by the author of the present volume (the *Times*, March 26). In it examples of drawings illustrative of the fading of indigo, and of the various cochineal pigments, were cited, while it was pointed out that many of the South Kensington drawings had changed for the worse long before their exhibition in the museum. The effects of direct sunshine, of diffused daylight, of gas and of hygroscopic moisture were named.

The use of a selected, restricted and perfectly safe, yet adequate palette was recommended. Finally, the real cause of the 'hot' appearance of skies and clouds represented by mixtures of indigo and iron-reds was explained.

Mr. Robinson returns to the charge (on March 26), expanding and reiterating his first statements. Then Sir J. D. Linton (April 3) once more combats Mr. Robinson's views, conceding, however, that changes have taken place in some of the earlier works by Girtin, Barret, De Wint, Havell, and others from 'exposure to damp, smoke, and direct sunlight, but chiefly from the use of Indian red,' these changes being stated to have occurred prior to the acquisition of the drawings by the museum. Sir J. D. Linton concludes by denying that he had spoken of chemical changes caused by the use of Indian red. [These, however, were his words, see *ante*, p. 315.]

The communications which appeared in the *Times* of April 3 and 14, from the pens of Mr. W. Severn, Mr. G. C. Bentinck, an Artist, and Mr. Ruskin, do not demand notice in this place. The present writer's second letter (April 14) described the chief changes in washes of pigments as due to the combined influence of moisture, oxygen and light, and referred to several other related topics, which will be found fully discussed in the present volume. Other contributions to the controversy, by Mr. Frank Dillon, Mr. Ayscough Fawkes, 'F.R.C.S.,' and by the two disputants who opened the debate, will be found in the *Times* of April 14, 20, 28, and June 11 and 14.

After a lull of some weeks the discussion was resumed in the *Times* of July 5. A letter from Mr. Robinson appeared on July 16; in this the protection from fading afforded to crimson lake by a vacuum, and by the sifting of the solar rays by means of a screen of crimson glass were mentioned. The notorious and free use of this pigment by many water-colour painters of the first half of this century was commented on in connection with the observation that three days' sunshine sufficed to take out all the colour from a washed tint of this crimson lake. A letter from (the late) Dr. Percy to Mr. Robinson, printed with the above-named communication, states the important fact that while greys made with indigo and Indian red have been preserved unchanged for fifty-seven years in the dark, the indigo in the same association has wholly disappeared in other instances where drawings have been long exposed to solar light. Dr. Percy drew attention to the curious purple tint which some plate-glass containing manganese acquires by exposure to sunlight, citing this change as evidence of chemical alteration in a hard, solid and compact body.

On July 24 Colonel (now General) Donnelly announced the appointment of a committee consisting of eight artists and the keeper of the prints in the British Museum, to consider the question of the action of light on paintings in water-colours. Mr. Robinson criticised the composition of this committee in the *Times* of July 28, and observed that measures had at last been taken to shield the South Kensington water-colours from the direct sunlight which was wont to impinge upon them. Colonel Donnelly rejoined 'that the measures now taken to protect the pictures from the direct action of sunlight are precisely the same as have been employed for many years past.'

Mr. Ruskin's strongly expressed views on the proper method of securing water-colour drawings from the injurious effects of light, and of other hostile influences, are quoted by Mr. Robinson in his letter, published in *The Times* of August 4. The passages cited are from the first volume of 'Arrows of the Chase,' pp. 129, 130, 149. The discussion was continued in *The Times* of August 10, where letters by Messrs. E. A. Goodall, F. Dillon, L. Fagan, J. C. Robinson, and James Orrock, appeared. Again, on the 11th, when Colonel Donnelly once more denied Mr. Robinson's assertions regarding the use made of the skylight blinds at South Kensington. Again, on the 17th, by Mr. Robinson; on September 6, by Mr. Orrock; and, on the 13th, by Mr. Robinson. These last three letters were mainly devoted to the nature, treatment, and condition of drawings belonging to Dr. Percy and Mr. Orrock. A valuable communication from Mr. W. Simpson also appeared in *The Times* of September 13, 1886, but, as it referred to actual trials of pigments, its conclusions have been already reported (page 305).

Other contributions to this discussion remain to be noticed. These appeared in *The Athenæum*, *The Saturday Review*, *The Nineteenth Century*, and *The Art Journal.*

The Athenæum (June 12, 1886) is of opinion that a committee of painters drawn from the Royal Water-Colour Society 'would not sanction such flagrantly absurd statements as that because Egyptian distemper pictures exist of any age, therefore all kinds of water-colour drawings are permanent; or, on the other hand, that because water-colour drawings, after having been exposed to strong light for lengthened periods of time, have more or less faded, therefore the art cannot be relied on. Our experience justifies the opinion that nine out of ten, if long exposed to strong light, suffer more or less, some being completely ruined, others scarcely affected at all. A few escape altogether, but these are of the nature of the Egyptian distempers, and for very long periods may be trusted to defy the sun himself. Broadly speaking, it is safest not continuously to expose drawings to strong light; but, on the whole, daily experience declares that vast numbers of drawings have been hung in rooms illuminated in the ordinary manner, and have suffered in no considerable degree, or very little indeed. The truth lies between the extremes of what has been said concerning the drawings now at South Kensington. A very few are practically intact; the majority have suffered by the way in which they are shown to the public, a way which ensures the destruction of everything not executed in the manner of the Egyptians.'

In an article on 'The Preservation of Water-Colours,' which appeared in *The Saturday Review* of July 3, 1886, the complex and varied nature of the changes generally called 'fading' is emphasized. The rigorous protection from pure sunlight of water-colour drawings is urged, while it is recommended that exposure to ordinary daylight should be limited so far as may be possible. It is affirmed that 'Mr. Robinson obviously and grotesquely overshoots his mark when he asserts that "all" the drawings in the South Kensington Museum have been "more or less irreparably injured." Sir James Linton writes still more wildly when he avers that those collected in his exhibition "are all in a perfect condition, brilliant, unfaded, beyond the contradiction of even Mr. Church or Mr. Robinson. Mr. Ruskin joins in the affray with his authoritative air and his audacity of fence, and is, as usual, incisive and provocative, felicitously right and preposterously wrong in a breath.'

In connection with the above discussion I may refer to an important paper which appeared (before its commencement) in the *Nineteenth Century* for March, 1886. In it Mr. W. G. Rawlinson, speaking of the early English water-colourists, says, 'Time has dealt hardly with the majority of their pictures, and few of even the more important drawings that now come into the market but have faded more or less. I do not hesitate to say that only those who have the good fortune to know collections such as that of Dr. Percy and the one in the Print Room of the British Museum (which are habitually guarded from light and kept in portfolios) can form an adequate idea of the delicacy, force, and beauty of colour which were attained by the artists I have named, and by others of their less known contemporaries. A large proportion of the splendid Ellison and Townshend bequests permanently exhibited at South Kensington have greatly deteriorated, and we have only to look at several of Turner's drawings in the present Exhibition' (R. A., January, 1886), 'to see the havoc which undue exposure has made of them. . . . . The avoidance of anything like direct sunlight is an absolute necessity.'

In *The Art Journal* for August, 1886 (page 252), Mr. Walter Armstrong makes the assertion that the author of the present handbook 'declares practically that all water-colour drawings begin to die away on exposure to ordinary daylight.' I have endeavoured, without success, to obtain from Mr. Armstrong a reference to any words of mine which could bear such an interpretation.

The last communication on this subject which I mention here is an article by Mr. Frank Dillon, in *The Nineteenth Century* of August, 1886. It extends over nine pages, and contains many data of importance.

INDEX

BILLING AND SONS, PRINTERS, GUILDFORD.

Spotted something that isn't quite right?

Let us know and we'll fix it for future reprints. And as a thank you for your diligence, we'll send you a free Sothis book of your choice.

publisher@sothis.press

This page intentionally left blank

Made in United States
Orlando, FL
05 November 2026

90231871R00199